Radioactive Isotopes in Biological Research

Radioactive Isotopes in Biological Research

WILLIAM R. HENDEE

Associate Professor of Radiology
University of Colorado Medical Center
Denver, Colorado

A Wiley-Interscience Publication
JOHN WILEY & SONS
New York   London   Sydney   Toronto

Library of Congress Cataloging in Publication Data

Hendee, William R   1938–
    Radioactive isotopes in biological research.

    "A Wiley-Interscience publication."
    Includes bibliographical references.
    1. Radioisotopes in biology.   I. Title.
    [DNLM: 1. Biology.   2. Radioisotopes.
    3. Research.   WN415 H495r 1973]
    QH652.H38   574.1'915   73-8966
    ISBN 0-471-37043-6

Printed in the United States of America

10 9 8 7 6 5 4 3 2 1

To Mom and Dad

Preface

The rapid growth of knowledge in the fields of biochemistry, cellular physiology, and molecular biology reflects to a considerable degree the utilization of radioactive isotopes as tracers of chemical and biological processes. The emergence of nuclear medicine as a separate medical specialty reflects the current importance of radioactive materials in the biological and medical sciences, and the almost unlimited potential of radioimmunoassay analysis suggests that utilization of radioactive isotopes will continue to stimulate the growth of biomedical knowledge. Because of the current and anticipated importance of radioisotopes in the biological and medical sciences, students working toward careers in these fields may be well-advised to participate in a course in radioisotope techniques. Although the course might be taken as late as the first year or two of graduate study, its inclusion in the undergraduate curriculum may be preferable. It is for such a course, either graduate or undergraduate, that this text is written. It is hoped that the text may prove useful also to researchers using radioactive isotopes without formal training in radioisotope techniques.

This text has evolved from experience in teaching a course in radioisotope techniques to undergraduate and graduate students in the biological, chemical, and medical sciences. To these students, whose questions, interest, and willingness to spend long hours in the laboratory have been a continual stimulus for my own teaching and research, my appreciation and gratitude is offered. What is presented well in this book is a reflection of the influence of present and former students upon my growth as a teacher. What is presented poorly is an indication of my need for continued growth.

To Paul L. Carson, Edward L. Chaney, and Elaine J. Cuklanz, all of the Medical Physics Section at the University of Colorado Medical Center, I extend my appreciation for setting my thoughts straight on certain technical matters which were in a rather disarrayed state in my own mind. My sincere thanks are offered also to Miss Carolyn Yandle and Mrs. Elizabeth Case for typing the manuscript many times over, and to

Mrs. Carolyn Finster for typing assistance during my more ambitious periods of writing. I am grateful to Geoffrey Ibbott, James Burns, and Dale Fitting for the accumulation of data for many of the figures. Illustrations were prepared by Mrs. Josephine Ibbott and Mr. Ken Crusha, who adjusted with stoic forebearance to an erratic schedule of frantic activity separated by weeks of inactivity.

Foremost of all, I owe another debt of gratitude to my wife Jeannie and our seven children, all of whom supported and encouraged the development of yet another book.

Denver, Colorado   *William R. Hendee*

Contents

Radioactive Isotopes in Biological Research

1

Structure
of the
Atom

An atom is composed of a small, positively charged nucleus surrounded by a cloud of negatively charged electrons. Each electron has a mass of 9.1×10^{-28} g and possesses one electronic charge of negative sign, equal to -1.6×10^{-19} coul. The electrons are confined to orbits at prescribed distances from the nucleus, where they revolve with speeds adequate to cause them to remain in their orbits and not spiral into the nucleus. So long as an electron remains in a particular orbit, it does not lose or gain energy. If the electron moves to an orbit closer to the nucleus, energy is released; energy must be supplied for an electron to move to an orbit farther from the nucleus, because work must be done against the attractive force of the nucleus.

The model of the atom described above was developed in 1913 by Niels Bohr, and is described as the Bohr model of the atom. Although more sophisticated models of the atom have been developed, the Bohr model is sufficient to explain properties of the atom at the level of this text.

Electron Orbits and Electron Binding Energy

An electron in an inner orbit of an atom is attracted to the positively charged nucleus with a force greater than that exerted by the nucleus upon an electron farther away. The inner electron may move to an orbit farther from the nucleus, or to a position outside the atom, only if energy

1

is supplied for the transition. The energy required to remove an electron completely from an atom is the binding energy E_B for the electron. The binding energy is greater numerically for an electron close to the nucleus than for an electron farther away, because the attractive force of the nucleus is greater for the inner electron and more energy is required to remove it from the atom. The energy required to move an electron from one orbit to another orbit more distant from the nucleus is the difference in binding energy of the electron in the two orbits.

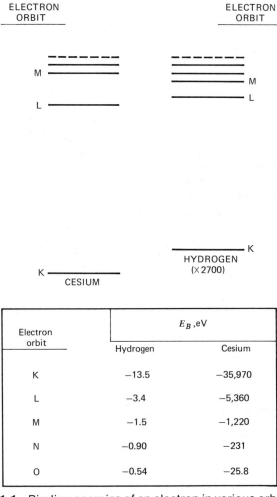

Electron orbit	E_B ,eV	
	Hydrogen	Cesium
K	−13.5	−35,970
L	−3.4	−5,360
M	−1.5	−1,220
N	−0.90	−231
O	−0.54	−25.8

Figure 1-1   Binding energies of an electron in various orbits of hydrogen ($Z = 1$) and cesium ($Z = 55$).

The binding energy of electrons in a particular orbit increases with the positive charge of the nucleus. In Figure 1-1, binding energies are compared for electrons in hydrogen (charge of $+1$ for the nucleus) and cesium (charge of $+55$ for the nucleus). The energy required to move an electron from the innermost electron orbit (K shell) to the next orbit (L shell) is

$$[-3.4\,eV - (-13.5\,eV)] = 10.1\,eV^* \text{ in hydrogen}$$

$$[-5360\,eV - (-35{,}970\,eV)] = 30{,}610\,eV \text{ in cesium}$$

where $-13.5\,eV$ and $-35{,}970\,eV$ are the binding energies of an electron in the K shell of hydrogen and cesium, respectively, and $-3.4\,eV$ and $-5360\,eV$ are the binding energies of an electron in the L shell of the same elements.

Electron Transitions and Characteristic Radiation

When an electron is removed from an orbit near the nucleus, a vacancy or "hole" is created in the orbit. This vacancy is filled promptly by transition of an electron from an orbit farther from the nucleus. During transition of the electron, an amount of energy is released equal to the difference in binding energy of the electron in its original and final orbit. For example, $[-35{,}970\,eV - (-1220\,eV)] = -34{,}750\,eV$ of energy are released when an electron moves from the M to the K orbit in cesium (Figure 1-1). The negative sign implies the release of energy. Usually, the energy is released as a photon or "packet" of electromagnetic radiation referred to as characteristic radiation, because the energy is characteristic of the difference in binding energy of an electron in two orbits of a particular atom. If the energy of the characteristic radiation exceeds 100 eV, the radiation is classified as x-radiation; if the energy is

*An electron volt is a unit of energy defined as the kinetic energy of an electron accelerated through a potential difference of 1 volt *in vacuo*. 1 eV = $1\cdot6 \times 10^{-19}$ joule (J); 1 keV = 10^3 eV; 1 MeV = 10^6 eV.

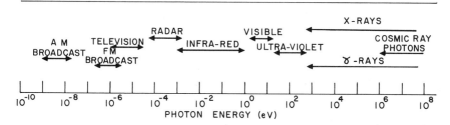

Figure 1-2   The energy spectrum of electromagnetic radiation.

less than 100 eV, the radiation is termed ultraviolet, visible, or infrared radiation (Figure 1-2). Depending upon the orbit assumed by the cascading electron, the characteristic radiation may be described as K-characteristic radiation or K-fluorescence, L-characteristic radiation or L-fluorescence, etc.

Auger Electrons and Fluorescence Yields

An electron moving from one orbit to another closer to the nucleus does not always initiate the release of electromagnetic radiation. Instead, the energy sometimes is transferred to another electron which is ejected from the atom. The ejected electron, termed an *Auger electron*, usually originates from the same orbit as the cascading electron, and possesses a kinetic energy equal to the energy released during transition of the cascading electron, less the binding energy of the ejected electron (Figure 1-3).

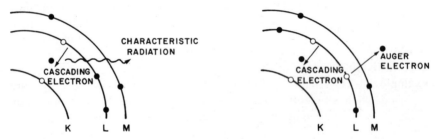

Figure 1-3 An electron transition from one orbit to another nearer the nucleus is accompanied by either the emission of characteristic radiation (left) or the release of an Auger electron (right).

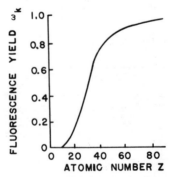

Figure 1-4 Fluorescence yield ω_k plotted as a function of atomic number Z.

The probability that transition of an electron into a particular electron orbit will be accompanied by emission of characteristic radiation is termed the fluorescence yield ω for the orbit. The probability is $1 - \omega$ that the transition will result in the ejection of an Auger electron. The fluorescence yield ω_k for transition of an electron into the K orbit is plotted in Figure 1-4 as a function of the charge of the nucleus (atomic number Z) [1]. The fluorescence yield increases with increasing charge of the nucleus, reflecting the greater binding of electrons to nuclei with greater charge.

Models and Nomenclature of the Nucleus

In a neutral atom, the nucleus possesses a positive charge equal in magnitude but opposite in sign to the total charge of all electrons outside the nucleus. If the atom contains Z electrons, then the total negative charge of all the electrons is $Z \times (-1.6 \times 10^{-19}$ coul). Consequently, the positive charge of the nucleus also is $Z \times (+1.6 \times 10^{-19}$ coul). The positive charge of the nucleus, and roughly half of its mass, are contributed by protons. Each proton has a charge of $+1.6 \times 10^{-19}$ coul, and there are as many protons inside the nucleus as there are electrons outside. The number of protons (or electrons) is referred to as the atomic number Z of the atom.

The mass of a proton is 1.6724×10^{-24} g. The remainder of the nuclear mass is contributed by neutrons, neutral particles with a mass of 1.6747×10^{-24} g. The number of neutrons in a particular nucleus is termed the *neutron number N* for the nucleus. Neutrons and protons often are referred to collectively as nucleons, and the number of nucleons in a nucleus is termed the *mass number A* for the nucleus (that is, $A = N + Z$). Neutrons outside the nucleus are unstable, decaying into a proton, an electron and a massless neutral particle termed a neutrino. The half-life of this transition is 12.8 m.*

Each element has a characteristic number of protons and electrons. Consequently, an element is defined explicitly by its atomic number Z. Isotopes of a particular element are atoms which have the same number of protons but different numbers of neutrons. A particular isotope is designated by the chemical symbol for the element together with the mass number for the isotope as a left superscript. For example,

$$^1H, {}^2H, {}^3H$$

---

*The half-life is the time required for half of a sample of items (for example, neutrons) to decay.

are isotopes of the element hydrogen. Protium (^{1}H) possesses one proton and no neutrons, deuterium (^{2}H) has one proton and one neutron, and tritium (^{3}H) contains one proton and two neutrons. Sometimes the atomic number is added to this designation as a left subscript:

$$^1_1\text{H}, {}^2_1\text{H}, {}^3_1\text{H}$$

As another example,

$$^9\text{C}, {}^{10}\text{C}, {}^{11}\text{C}, {}^{12}\text{C}, {}^{13}\text{C}, {}^{14}\text{C}, {}^{15}\text{C}, {}^{16}\text{C}$$

are isotopes of the element carbon, with the number of neutrons ranging from 3(^{9}C) to 10(^{16}C). A nucleus of carbon always contains 6 protons.

Isotones are atoms which contain the same number of neutrons but different numbers of protons. For example,

$$^8\text{He}, {}^9\text{Li}, {}^{10}\text{Be}, {}^{11}\text{B}, {}^{12}\text{C}, {}^{13}\text{N}, {}^{14}\text{O}$$

are isotones, because each atom contains 6 neutrons. Isobars are atoms which contain the same number of nucleons (protons plus neutrons). For example,

$$^{13}\text{B}, {}^{13}\text{C}, {}^{13}\text{N}, {}^{13}\text{O}$$

are isobars because each contains 13 nucleons. Isomers are different energy states of nuclei with the same number of neutrons and the same number of protons. A nuclide is a nucleus in any form.

Term	Atomic number Z	Neutron number N	Mass number A
Isotopes	Same	Different	Different
Isotones	Different	Same	Different
Isobars	Different	Different	Same
Isomers	Same	Same	Same, (different nuclear energy states)

Nuclear Force and Nuclear Binding Energy

An electrostatic repulsive force exists between particles with a similar charge, and protons repel each other when separated by a distance greater than the diameter of the nucleus. Within the nucleus, however, protons remain together. Consequently, a force must exist within the dimensions of the nucleus ($\sim 10^{-13}$ cm) which counteracts the electrostatic repulsive force. This attractive force, termed the *nuclear* (or strong) *force*, is about 100 times greater than the electrostatic force, and is responsible for holding neutrons and protons together within the nucleus. The nuclear force is effective only over distances less than about 10^{-12} cm.

Compared to conventional units of mass, the masses of atomic particles are very small and usually are described in terms of the atomic mass unit. The atomic mass unit (amu) is defined as

$$1 \text{ amu} = 1/12 \text{ of the mass of an atom of } {}^{12}\text{C}$$

One amu is equivalent to 1.6605×10^{-24} g. Masses of atomic particles expressed in amu's are:

$$\text{electron} = 0.00055 \text{ amu}$$
$$\text{proton} = 1.00727 \text{ amu}$$
$$\text{neutron} = 1.00866 \text{ amu}$$

The isotope ${}^{14}\text{N}$ has 7 protons, 7 neutrons, and 7 electrons. The total mass of the components of an atom of ${}^{14}\text{N}$ is:

mass of 7 protons	$= 7(1.00727 \text{ amu}) =$	7.05089 amu
mass of 7 neutrons	$= 7(1.00866 \text{ amu}) =$	7.06062 amu
mass of 7 electrons	$= 7(0.00055 \text{ amu}) =$	0.00385 amu
mass of components of ${}^{14}\text{N}$	$=$	14.11536 amu

The actual mass of the ${}^{14}\text{N}$ atom has been determined to be 14.00307 amu[2]. The difference between the mass of the components of an atom and the actual mass of the atom is termed the *mass defect*. For ${}^{14}\text{N}$,

$$\text{mass defect} = 14.11536 \text{ amu} - 14.00307 \text{ amu}$$
$$= 0.11229 \text{ amu}$$

According to the formula for mass-energy equivalence proposed in 1905 by Einstein[3], the mass defect of an atom represents the energy which must be supplied to separate the atom into its component neutrons, protons and electrons. Mass-energy equivalence may be stated as

$$E = mc^2$$

where E represents the energy equivalent to mass m, and c is a coefficient equal numerically to the speed of light in vacuo (3×10^{10} cm/sec). For example, one kilogram of mass is equivalent to 9×10^{16} joule, an amount of energy equal roughly to that released during detonation of 30 million tons of TNT. The energy equivalent to one amu is:

$$\frac{(1 \text{ amu}) (1.66 \times 10^{-24} \text{ g/amu})(3 \times 10^{10} \text{ cm/sec})^2}{(1.6 \times 10^{-6} \text{ erg/MeV})} = 931 \text{ MeV}$$

The energy required to separate an atom of ${}^{14}\text{N}$ (mass defect = 0.11229 amu) into its components is:

$$(0.11229 \text{ amu})(931 \text{ MeV/amu}) = 104.5 \text{ MeV}$$

If the very small contribution of electrons to the mass defect is ignored, then the energy requirement of 104.5 MeV may be considered the nuclear binding energy for ^{14}N. The quotient of the nuclear binding energy divided by the number of nucleons in the nucleus is the average binding energy per nucleon. For ^{14}N, the average binding energy per nucleon (E_{Bavg}) is:

$$E_{Bavg} = (104.5 \text{ MeV})/(14 \text{ nucleons})$$

$$= 7.46 \text{ MeV/nucleon}$$

In Figure 1-5, the average binding energy per nucleon (in units of million electron volts per nucleon) is plotted for nuclei with mass numbers up to 240.

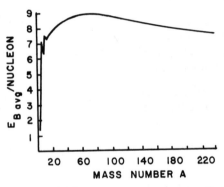

Figure 1-5 Average binding energy per nucleon E_{Bavg} in units of million electron volts per nucleon for nuclei with mass numbers from 1 to 240.

Example 1-1
Compute the average binding energy per nucleon for ^{4}He.

$$\begin{aligned}
\text{mass of 2 protons} &= 2(1.00727 \text{ amu}) = 2.01454 \text{ amu} \\
\text{mass of 2 neutrons} &= 2(1.00866 \text{ amu}) = 2.01732 \text{ amu} \\
\text{mass of 2 electrons} &= 2(0.00055 \text{ amu}) = \underline{0.00110 \text{ amu}} \\
\text{mass of components of } ^{4}\text{He} &= 4.03296 \text{ amu} \\
\text{mass of } ^{4}\text{He atom} &= 4.00260 \text{ amu} \\
\text{mass defect} &= 0.03036 \text{ amu}
\end{aligned}$$

If the small contribution of electrons to the mass defect of ^{4}He is ignored, then:

nuclear binding energy of ^{4}He $=$ (0.03036 amu)(931 MeV/amu)
$$= 28.27 \text{ MeV}$$

The average binding energy per nucleon for ^{4}He is:

$E_{Bavg} =$ (28.27 MeV)/4 nucleon
$$= 7.07 \text{ MeV/nucleon}$$

Nuclear Fission and Fusion

If a nucleus with high mass number separates or fissions into two parts, each with an average binding energy per nucleon greater than that for the original nucleus, then the total nuclear binding energy is increased and energy is released. Certain heavy nuclei (for example, ^{235}U and ^{239}Pu) fission spontaneously after absorbing a slowly moving neutron. For ^{235}U, a typical fission reaction is

$$^{235}_{92}\text{U} + ^1_0\text{n} \rightarrow ^{236}_{92}\text{U} \rightarrow ^{140}_{54}\text{Xe} + ^{93}_{38}\text{Sr} + 3^1_0\text{n}$$

The nuclides $^{140}_{54}$Xe and $^{93}_{38}$Sr are radioactive, and decay by successive β transitions to the stable nuclides $^{140}_{54}$Ce and $^{93}_{41}$Nb. The energy released during fission and during the subsequent β transitions may be computed as shown in Example 1-2. Four β particles, each with a mass of 0.00055 amu, are released during transition of ^{140}Xe to ^{140}Ce. Three β particles are released during transition of ^{93}Sr to ^{93}Nb. Four electrons are acquired by the atom during transition of ^{140}Xe to ^{140}Ce, and three electrons are acquired during transition of ^{93}Sr to ^{93}Nb.

Example 1-2

Compute the energy released during transition of ^{236}U to the stable products ^{140}Ce and ^{93}Nb.

mass before fission

mass of ^{236}U atom	$= 236.0525$ amu
mass of 7 electrons	$=0.0039$ amu
total	$= 236.0564$ amu

mass after fission and decay

mass of ^{140}Ce atom	$= 139.9053$ amu
mass of ^{93}Nb atom	$=92.9060$ amu
mass of 3 neutrons	$=3.0260$ amu
mass of 7 β particles	$=0.0039$ amu
total	$= 235.8412$ amu
mass difference	$= 0.2152$ amu
energy released	$= (0.2152 \text{ amu})(931 \text{ MeV/amu})$
	$= 200 \text{ MeV}$

A fission reaction results in the release of about 200 MeV of energy. This energy is released primarily as kinetic energy of fission products and neutrons, ionizing and nonionizing electromagnetic radiation, and as energy accompanying the decay of the product nuclei.

Products of fission such as ^{140}Xe and ^{93}Sr, and the nuclides formed as these products decay, are known as fission products. Many fission products are radioactive and are used in biologic research and medicine.

When a heavy nucleus fissions, the two or three neutrons released may be captured by other heavy nuclei. These nuclei also may fission and release more neutrons. Consequently, a chain reaction is possible, provided that a sufficient mass (critical mass) of fissionable material is contained within a small volume. The rate at which nuclei fission may be controlled by regulating the number of neutrons available each instant for absorption by heavy nuclei. This procedure is used to control the rate at which uranium or plutonium nuclei fission in a nuclear reactor. Uncontrolled fission can result in a nuclear explosion.

Energy is released also if two light nuclei are combined to form a nucleus with higher average binding energy per nucleon. This process, termed *nuclear fusion*, is illustrated by the following examples:

$$_1^2H + _1^2H \rightarrow _2^3He + _0^1n + 3.26 \text{ MeV}$$
$$_1^2H + _1^2H \rightarrow _1^3H + _1^1p + 4.04 \text{ MeV}$$
$$_1^2H + _1^3H \rightarrow _2^4He + _0^1n + 17.6 \text{ MeV}$$
$$_1^2H + _2^3He \rightarrow _2^4He + _1^1p + 18.3 \text{ MeV}$$

The energy released during a typical fusion reaction is computed as shown in Example 1-3.

Example 1-3

Neglecting the small influence of electrons on the energy released during fusion, estimate the energy released during the reaction:

$$^2H + {^3He} \rightarrow {^4He} + {^1p}$$

mass of ^{2}H	= 2.0141 amu
mass of ^{3}He	= 3.0160 amu
total mass before fusion	= 5.0301 amu

mass of ^{4}He	= 4.0026 amu
mass of ^{1}p	= 1.0078 amu (mass of ^{1}H used to balance electrons)
total mass after fusion	= 5.0104 amu

mass difference	= 5.0301 amu − 5.0104 amu
	= 0.0197 amu
energy released	= (0.0197 amu) (931 MeV/amu)
	= 18.3 MeV

For fusion to occur, two positively charged nuclei must be brought sufficiently near one another that the nuclear force can overcome the electrostatic repulsive force and initiate fusion. To attain this proximity, the nuclei must approach each other at very high velocities. Adequate velocities are attained only at temperatures of 10^6–10^{7}°K, equivalent to the temperature of the inner region of the sun. At the present time, temperatures this high are achieved on earth only at the center of a fission explosion. For this reason, a fusion (hydrogen) bomb must be "triggered" with a fission bomb. Methods for achieving controlled nuclear fusion (for example, by forcing the particles together by application of an intense magnetic field) have so far been unsuccessful.

PROBLEMS

1. Using the data in Figure 1-1, compute the energy required to move an electron from the K to the M orbit in cesium. How much energy is released when an electron moves from the M to the L orbit in the same element?

2. Group the following nuclides into isotopes, isotones and isobars:

$$^{45}_{21}\text{Sc}, \, ^{46}_{24}\text{Cr}, \, ^{45}_{23}\text{V}, \, ^{46}_{22}\text{Ti}, \, ^{47}_{22}\text{Ti}, \, ^{45}_{20}\text{Ca}, \, ^{44}_{22}\text{Ti}, \, ^{46}_{23}\text{V}, \, ^{45}_{22}\text{Ti}$$

3. Compute the mass defect and the average binding energy per nucleon (ignoring electron contribution) for ^{16}O. The mass of an atom of ^{16}O is 15.9949 amu. What are the atomic and mass numbers of ^{16}O?

4. Compute the energy released during the fission reaction and subsequent beta transitions described below:

$$^{235}_{92}\text{U} + ^{1}_{0}\text{n} \rightarrow \, ^{236}_{92}\text{U} \rightarrow \, ^{141}_{56}\text{Ba} + ^{92}_{36}\text{Kr} + 3^{1}_{0}\text{n}$$

$^{141}_{56}\text{Ba} \, 3 \, \beta \text{ transitions} \underset{\rightarrow \, \rightarrow \, \rightarrow}{} \, ^{141}_{59}\text{Pd} \quad \text{mass } ^{141}\text{Pd} = 140.9074 \text{ amu}$

$^{92}_{36}\text{Kr} \, 4 \, \beta \text{ transitions} \underset{\rightarrow \, \rightarrow \, \rightarrow \, \rightarrow}{} \, ^{92}_{40}\text{Zr} \quad \text{mass } ^{92}\text{Zr} = 91.9046 \text{ amu}$

5. Ignoring the contribution of electrons, estimate the energy released in the fusion reaction

$$^{2}\text{H} + ^{3}\text{H} \rightarrow \, ^{4}\text{He} + ^{1}\text{n}$$
$$\text{mass } ^{2}\text{H} = 2.0141 \text{ amu}$$
$$\text{mass } ^{3}\text{H} = 3.0160 \text{ amu}$$
$$\text{mass } ^{4}\text{He} = 4.0026 \text{ amu}$$
$$\text{mass } ^{1}\text{n} = 1.0086 \text{ amu}$$
$$\text{mass e}^{-} = 0.0005 \text{ amu}$$

6. The energy released during the nuclear explosion at Hiroshima was about equal to the energy released by detonation of 200,000 tons of TNT. If the energy released is about 200 MeV during fission of a ^{236}U nucleus and 3.8×10^9 J during detonation of 1 ton of TNT, compute the number of nuclei which fissioned in the Hiroshima explosion and estimate the total decrease in mass (1 MeV $= 1.6 \times 10^{-13}$ J).

REFERENCES

[1] Broyles, C., D. Thomas, and S. Haynes. 1953. The measurement and interpretation of the K Auger intensities of Sn^{113}, Cs^{137}, and Au^{198}. Phys. Rev. 89:715.
[2] Weast, R. C. (ed.). 1967–1968. Handbook of chemistry and physics, 48th Ed. The Chemical Rubber Co. Cleveland, Ohio.
[3] Einstein, A. 1905. Über einen die Erzeugung and Verwandlung des Lichtes betreffenden heuristischen Gesichtspunkt. Ann. Physik 17:132.

2
Radioactive Decay

In general, the total number of nucleons and the ratio of neutrons to protons determine whether a nucleus is stable or unstable. Unstable nuclei undergo nuclear transitions such as nuclear fission or, more frequently, radioactive decay. All isotopes of elements with atomic number greater than 83 are radioactive; in addition, many radioactive isotopes have been identified with atomic number less than 83. Radioactive isotopes (or nuclides) undergo nuclear transitions which usually are accompanied by emission of particulate and electromagnetic radiation.

Nuclei with odd numbers of neutrons and protons tend to be unstable, whereas nuclei with even numbers of neutrons and protons more frequently are stable. The number of stable isotopes identified for different combinations of neutrons and protons is shown below:

Number of protons, Z	Number of neutrons, N	Number of stable isotopes
Even	Even	165
Even	Odd	57
Odd	Even	53
Odd	Odd	6

In stable nuclei with low atomic numbers, the number of neutrons is about equal to the number of protons. In stable nuclei of higher Z, the number of neutrons exceeds the number of protons. This relationship between neutrons and protons in stable nuclei is illustrated in Figure 2-1.

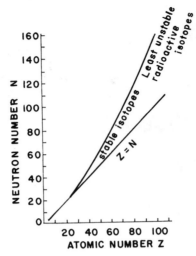

Figure 2-1 Number of neutrons in stable nuclei as a function of the number of protons (atomic number Z). No stable isotopes exist for elements with $Z > 83$.

Negatron Decay

A nucleus with a ratio of neutrons to protons too great for stability undergoes a nuclear transition which reduces the neutron:proton ($N:Z$) ratio. The transition involves transformation of a neutron ($_0^1n$) into a proton ($_1^1p$), a negative electron ($_{-1}^0e$), and a massless neutral particle named a *neutrino* (v).

$$_0^1n \rightarrow {_1^1}p + {_{-1}^0}e + v$$

The proton is retained in the nucleus, and the neutrino and negative electron (termed a *negatron*) are ejected. Frequently, this transition or radioactive decay process is written

$$_0^1n \rightarrow {_1^1}p + {_{-1}^0}\beta + v$$

where $_{-1}^0\beta$ indicates the nuclear origin of the ejected electron. The resulting "daughter" nucleus has one more proton, but the same total number of nucleons, as the original "parent" nucleus. Consequently, this decay process, referred to as β decay, or negatron decay, provides an increase of 1 in Z with no change in A. Typical negatron decay processes include:

$$_1^3H \rightarrow {_2^3}He + {_{-1}^0}\beta + v \qquad {_{15}^{32}}P \rightarrow {_{16}^{32}}S + {_{-1}^0}\beta + v$$

$$_6^{14}C \rightarrow {_7^{14}}N + {_{-1}^0}\beta + v \qquad {_{16}^{35}}S \rightarrow {_{17}^{35}}Cl + {_{-1}^0}\beta + v$$

Radioactive decay processes often are described by decay schemes, in which the downward direction of the transition reflects the release of energy during the decay process. For a particular radioactive isotope, the total energy released during decay is constant and is termed the *transition energy* for decay of the isotope. If the transition is drawn toward the right as well as downward in the decay scheme, then the number of protons—that is, the atomic number Z—is increased by the transition, and the decay process is negatron decay. If the transition is drawn toward the left as well as downward in the decay scheme, then the atomic number is reduced by the transition, and the decay process is positron decay, electron capture, or α decay. Decay schemes for the negatron-emitting isotopes ^{3}H, ^{14}C, ^{32}P, and ^{35}S are shown in Figure 2-2.

Negatrons emitted during a decay process are characterized by a maximum energy E_{max}. Most negatrons from a radioactive sample have an energy less than E_{max} for the isotope in the sample. The average energy of the emitted negatrons is about $E_{max}/3$ (Figure 2-3). During decay of a

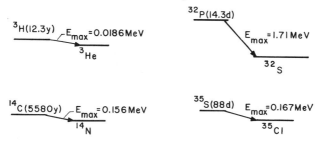

Figure 2-2   Decay schemes for negatron emitters ^{3}H, ^{14}C, ^{32}P, and ^{35}S.

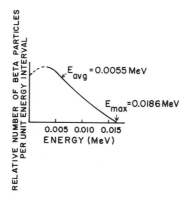

Figure 2-3   Energy spectrum for negatrons emitted by ^{3}H.

particular nucleus, the difference between E_{max} and the energy of the emitted negatron is removed from the nucleus by the neutrino. In the decay scheme for ^{3}H, for example,

$$\text{transition energy} = E_{max} = E_k + E_v$$

where E_k represents the kinetic energy of a particular negatron and E_v represents the energy associated with the accompanying neutrino. The maximum and average energies of a few common radioactive isotopes are listed in Table 2-1.

TABLE 2-1 *Maximum and Average Energies of a Few Common Radioactive Isotopes*

Nuclide	E_{avg}, MeV	E_{max}, MeV
^{3}H	0.0055	0.0186
^{14}C	0.050	0.156
^{22}Na	0.214	0.544
^{32}P	0.700	1.71
^{35}S	0.0492	0.167
^{45}Ca	0.077	0.256
^{65}Zn	0.143	0.325

Electron Capture and Positron Decay

A nucleus with a ratio of neutrons to protons too low for stability undergoes a transition which increases the $N:Z$ ratio. One possible transition is the capture of an orbital electron by the nucleus, resulting in the nuclear transformation

$$^1_1p + ^{\,0}_{-1}e \rightarrow ^1_0n + v$$

In most nuclei undergoing electron capture, about 90% of the electrons are captured from the K orbit (K capture) and about 10% from the L orbit (L capture). However, if the transition energy for the decay process is less than the binding energy of electrons in the K orbit, then electrons are captured only from the L and higher shells. If the transition energy exceeds the binding energy of a captured electron, then the difference in energy is removed by the neutrino and, sometimes, by γ-ray photons. Electron capture produces a vacancy in an inner electron orbit, and this mode of decay is accompanied by emission of characteristic radiation and Auger electrons.

A few representative transitions for decay by electron capture are:

$$^{51}_{24}\text{Cr} + {}_{-1}^{0}\text{e} \rightarrow {}^{51}_{23}\text{V} + v \qquad {}^{72}_{34}\text{Se} + {}_{-1}^{0}\text{e} \rightarrow {}^{72}_{33}\text{As} + v$$

$$^{54}_{25}\text{Mn} + {}_{-1}^{0}\text{e} \rightarrow {}^{54}_{24}\text{Cr} + v \qquad {}^{131}_{55}\text{Cs} + {}_{-1}^{0}\text{e} \rightarrow {}^{131}_{54}\text{Xe} + v \ .$$

A decay scheme for ^{51}Cr is shown in Figure 2-4.

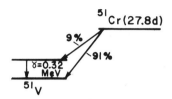

Figure 2-4   Scheme for decay of ^{51}Cr by electron capture.

Nuclei with a ratio of neutrons to protons too low for stability also may decay by emission of a positron. In this mode of decay, a proton is transformed into a neutron, a positive electron (positron), and a neutrino.

$$^{1}_{1}\text{p} \rightarrow {}^{1}_{0}\text{n} + {}_{+1}^{0}\text{e} + v$$

$$^{1}_{1}\text{p} \rightarrow {}^{1}_{0}\text{n} + {}_{+1}^{0}\beta + v$$

where ${}_{+1}^{0}\beta$ indicates the nuclear origin of the ejected positron. A few transitions representative of positron decay are shown below.

$$^{30}_{15}\text{P} \rightarrow {}^{30}_{14}\text{Si} + {}_{+1}^{0}\beta + v \qquad {}^{15}_{8}\text{O} \rightarrow {}^{15}_{7}\text{N} + {}_{+1}^{0}\beta + v$$

$$^{10}_{6}\text{C} \rightarrow {}^{10}_{5}\text{B} + {}_{+1}^{0}\beta + v \qquad {}^{23}_{12}\text{Mg} \rightarrow {}^{23}_{11}\text{Na} + {}_{+1}^{0}\beta + v$$

The mass of the products of positron decay exceeds the mass of the original proton. The deficiency in mass must be supplied by energy released during the decay process. The energy requirement is 1.02 MeV. Isotopes with a neutron:proton ratio too low for stability, but with a transition energy of less than 1.02 MeV, decay only by electron capture. Isotopes with transition energies greater than 1.02 MeV may decay by either electron capture or positron decay. The electron capture branching ratio for these isotopes describes the probability of decay by electron capture. In isotopes which exhibit both modes of decay, positron emission is favored for those with smaller mass numbers, and electron capture is favored for the heavier isotopes. For example, 57% of ^{52}Fe nuclei decay by positron emission, and 43% decay by electron capture (Figure 2-5). For the heavy isotope ^{207}Po, more than 99% of the decay processes are electron capture, and less than 1% are positron decay. In Figure 2-5, the

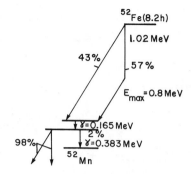

Figure 2-5 Scheme for decay of ^{52}Fe by electron capture and positron emission.

vertical portion of the pathway designating positron decay represents the portion (1.02 MeV) of the transition energy expressed as increased mass of the decay products.

Energy spectra for positrons resemble those for negatrons. The difference between E_{max} and the energy of a particular positron is carried away by a neutrino.

A few isotopes decay by both negatron decay and positron decay or electron capture. Primarily, these nuclides have odd numbers of neutrons and protons, and gain stability by either increasing or decreasing the atomic number. The multiple modes of decay for ^{64}Cu are shown in Figure 2-6.

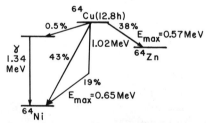

Figure 2-6 Decay scheme for ^{64}Cu, a nucleus with odd numbers of neutrons and protons, which decays by negatron decay, positron decay, and electron capture.

γ *Emission*

Radioactive decay of a nucleus often results in formation of a daughter nucleus in an excited unstable state. As the daughter nucleus changes from the excited state to a state of lower energy, energy is released, usually as electromagnetic radiation. The change is termed an *isomeric transition*, because the nucleus decays from one energy state to another with no change in *Z* or *A*. The state of least energy for the nucleus is termed the *ground state* or *ground energy level*. Typical γ transitions are included in the decay schemes for ^{64}Cu (Figure 2-6), ^{137}Cs (Figure 2-7), and ^{60}Co (Figure 2-8).

Occasionally, an isomeric state of a daughter nucleus may persist for a finite period before transition to a state of lower energy. If the half-life is greater than 10^{-6} sec for the isomeric state, then it is termed a *metastable*

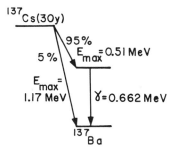

Figure 2-7 Decay scheme for ^{137}Cs, with 5% of the transitions directly to the ground state of ^{137}Ba and 95% of the transitions to an isomeric state 0.662 MeV about the ground state. Transition from the isomeric to the ground state is accompanied usually by the release of a γ-ray of 0.662 MeV.

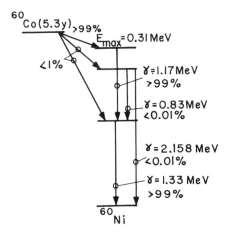

Figure 2-8 Decay scheme for ^{60}Co, which decays by negatron emission followed usually by isomeric transitions of 1.17 and 1.33 MeV.

state and described by a small "m" following the mass number (for example, ^{99m}Tc, ^{113m}In, ^{97m}Nb). A decay scheme for ^{24}Ne, including the metastable state ^{24m}Na for the daughter, is shown in Figure 2-9.

The energy E of a γ-ray (or x-ray) is related to its frequency v by the expression:

$$E = hv$$

where h is Planck's constant, 6.62×10^{-27} erg-sec. Electromagnetic radiation travels *in vacuo* (and, to a close approximation, in air) at a constant speed c of 3×10^{10} cm/sec. The speed of electromagnetic radiation is the product of the frequency v and wavelength $λ$.

$$c = vλ.$$

Consequently,

$$v = \frac{c}{λ}$$

and $E = hv = h\dfrac{c}{λ}$

The energy in keV of a photon of wavelength $λ$ in angstroms* is

$$E = \frac{12.4}{λ}$$

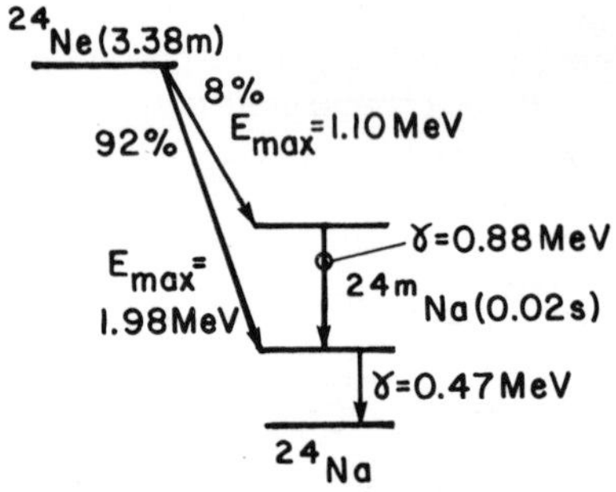

Figure 2-9 Decay scheme for ^{24}Ne, including the metastable state ^{24m}Na ($T_{1/2} = 0.02$ sec) for the daughter ^{24}Na.

Example 2-1
Derive the expression $E = 12.4/λ$.

$$E = h\frac{c}{λ} = \frac{(6.62 \times 10^{-27}\ \text{erg-sec})(3 \times 10^{10}\ \text{cm/sec})}{(10^{-8}\ \text{cm/Å})(1.6 \times 10^{-9}\ \text{erg/keV})}$$

$$E\,(\text{keV}) = \frac{12.4}{λ(\text{Å})}$$

*One angstrom (Å) equals 10^{-8} cm.

Example 2-2
Compute the energy of a γ-ray photon with a wavelength λ of 0.1 Å.

$$E = \frac{12.4}{\lambda}$$
$$= \frac{12.4}{0.1 \text{ Å}}$$
$$= 124 \text{ keV}$$

Internal Conversion

Usually, an isomeric transition of a nucleus is accompanied by the release of energy as electromagnetic radiation. Sometimes, however, the energy is transferred to an inner electron, and the electron is ejected from the atom with the energy released during the isomeric transition, reduced by the binding energy of the electron. The ejected electron is termed a *conversion electron* and the decay process is referred to as *internal conversion*. After ejection of the conversion electron, characteristic x-rays and Auger electrons are released as the extranuclear structure of the atom resumes a stable configuration.

The internal conversion coefficient α for a particular electron orbit in an atom is the ratio of conversion electrons from the orbit to the number of γ-rays from the nucleus. For example, the internal conversion coefficient α_K is 0.31 for the K orbit of ^{93m}Tc ($T_{1/2} = 43$ min), and 31 conversion electrons are ejected from the K shell for every 100 γ-rays emitted from the nucleus. The probability of internal conversion increases rapidly with the atomic number of the nucleus and the half-life of its excited state.

Example 2-3
How many conversion electrons are expected from the K shell during 100 isomeric transitions of ^{93m}Tc?

$$\text{number of conversion electrons} = \frac{\alpha_K}{1 + \alpha_K}(100)$$
$$= \frac{0.31}{1.31}(100)$$
$$= 24$$

If internal conversion is negligible for electrons in other orbits, then,

$$\text{number of } \gamma\text{-rays} = \frac{1}{1 + \alpha_K}(100)$$
$$= \frac{1}{1.31}(100)$$
$$= 76$$

α Decay

An additional mode of decay, involving the emission from the nucleus of two protons and two neutrons bound together as a helium nucleus ($^4_2$He), is available to some nuclei with high mass numbers. The emitted particle is named an α particle and the decay process is termed α decay. A typical α transition is:

$$^{226}_{88}\text{Ra} \rightarrow {}^{222}_{86}\text{Rn} + {}^4_2\text{He}$$

A decay scheme for α decay of ^{226}Ra is shown in Figure 2-10. The α particles from a particular nuclide are emitted with discrete energies. Rarely are α-emitting nuclides used in chemical and biological research, because the ability of α particles to penetrate matter is very limited, and because elements with α-emitting isotopes are present only in trace quantities in most living organisms.

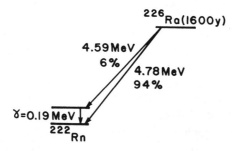

Figure 2-10   Decay scheme for α decay of ^{226}Ra.

Equations for Radioactive Decay

The rate of decay of a radioactive sample is described as the activity of the sample. Activity is described in units of disintegrations per second (dps), disintegrations per minute (dpm), or, more often, submultiples of the curie, where

$$
\begin{aligned}
1 \text{ curie (Ci)} \quad &= 3.7 \times 10^{10} \text{ dps}\\
1 \text{ millicurie (mCi)} \quad &= 10^{-3} \text{ Ci} = 3.7 \times 10^{7} \text{ dps}\\
1 \text{ microcurie } (\mu\text{Ci}) \quad &= 10^{-6} \text{ Ci} = 3.7 \times 10^{4} \text{ dps}\\
1 \text{ nanocurie (nCi)} \quad &= 10^{-9} \text{ Ci} = 37 \text{ dps}
\end{aligned}
$$

The activity of a sample varies directly with the number of radioactive atoms in the sample. This relationship is described by the expression

$$A = \lambda N$$

where A is the activity of the sample in units of disintegrations per unit time (for example, dps or dpm), N is the number of radioactive atoms present, and λ is a decay constant representing the fractional rate of decay of the sample in units of time^{-1}. The decay constant is related to the half-life $T_{1/2}$ of the sample by the expression

$$\lambda = \frac{\ln 2}{T_{1/2}} = \frac{0.693}{T_{1/2}}$$

where the half-life is the time required for decay of half of the radioactive atoms in the sample. This expression is derived in Appendix 1. Each radioactive isotope has a characteristic half-life and decay constant, as demonstrated in Table 2-2.

TABLE 2-2 *Half-Lives and Decay Constants for Common Radioactive Isotopes*

Nuclide	Decay Constant	Half-life
^{3}H	5.63×10^{-2}/year	12.3 years
^{14}C	1.24×10^{-4}/year	5580 years
^{32}P	4.84×10^{-2}/day	14.3 days
^{35}S	7.87×10^{-3}/day	88 days
^{36}Cl	2.23×10^{-6}/year	3.1×10^5 years
^{45}Ca	4.20×10^{-3}/day	165 days

If the activity A_0 of a radioactive sample is known at some time $t = 0$, then the activity A at a later time t may be computed with the expression

$$A = A_0 e^{-\lambda t}$$

Similarly, the number N of radioactive atoms in a sample at time t may be determined from

$$N = N_0 \, e^{-\lambda t}$$

where N_0 is the number of radioactive atoms present at an earlier time $t = 0$. The expressions for A and N are derived in Appendix 1. In appendix 2 are listed values for the constant e, known as the *exponential quantity*, raised to different positive and negative powers. The number N_d of atoms which have decayed in time t is

$$N_d = N_0 - N$$
$$= N_0 - N_0 e^{-\lambda t}$$
$$= N_0 (1 - e^{-\lambda t})$$

Example 2-1
Determine the activity of a sample of ^{32}P after five half-lives have elapsed, if the initial activity is 100 mCi.

$$A = A_0 e^{-\lambda t}$$
$$= A_0 e^{-(0.693/T_{1/2})t}$$
$$= 100 \text{ mCi} \left[e^{-(0.693/T_{1/2})(5T_{1/2})} \right]$$
$$A = 100 \text{ mCi } e^{-3.47}$$

From Appendix 2, $e^{-3.47} = 0.031$. Therefore,

$$A = 100 \text{ mCi } (0.031)$$
$$= 3.1 \text{ mCi}$$

Alternately, since $\frac{1}{2}A_0$ of the sample remains after one half-life,

$(\frac{1}{2})(\frac{1}{2}A_0) = (\frac{1}{2})^2 A_0 = \frac{1}{4}A_0$ remains after two half-lives

$(\frac{1}{2})^3 A_0 = \frac{1}{8}A_0$ remains after three half-lives

$(\frac{1}{2})^4 A_0 = \frac{1}{16}A_0$ remains after four half-lives

$(\frac{1}{2})^5 A_0 = \frac{1}{32}A_0 = 3.1$ mCi remains after five half-lives

Example 2-2
Determine the activity of a sample of ^{35}S after 110 days, if the initial activity is 25 mCi.

$$A = A_0 e^{-\lambda t}$$
$$= (25 \text{ mCi}) e^{-(0.693/T_{1/2})(110 \text{ days})}$$
$$= (25 \text{ mCi}) e^{-0.866}$$
$$= (25 \text{ mCi})(0.42)$$
$$A = 10.5 \text{ mCi}$$

The activity of a radioactive sample is illustrated in Figure 2-11 as a function of time expressed in multiples of the sample half-life. Since the logarithm of sample activity decreases linearly with time, data in Figure 2-11 may be replotted on semilogarithmic graph paper to yield a straight line (Figure 2-12).

Example 2-3
In 10 mg sample of $CaCO_3$, 50% of the carbon atoms are radioactive ^{14}C. How many ^{14}C atoms are present and what is the activity of the sample?

Half or 5 mg (0.005 g) of the $CaCO_3$ sample contains ^{14}C. Since the gram-molecular weight of $Ca^{14}CO_3$ is approximately 102, the mass of ^{14}C present is:

$$\text{mass } ^{14}\text{C} = (0.005 \text{ g}) \left(\frac{14}{102} \right)$$
$$= 6.86 \times 10^{-4} \text{ g}$$

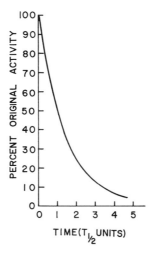

Figure 2-11   Activity of a radioactive sample as a function of time expressed in multiples of the sample half-life.

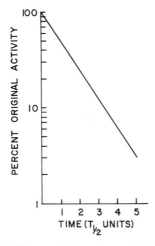

Figure 2-12   Data in Figure 2-11 replotted with activity on a semilogarithmic scale.

The number of atoms of ^{14}C is:

$$\text{atoms } ^{14}C = \frac{(\text{grams } ^{14}C)\; N_A}{M}$$

where N_A is Avogadro's number (number of atoms per gram-isotopic mass), and M is the gram-isotopic mass for ^{14}C.

$$\text{atoms } {}^{14}C = \frac{(6.86 \times 10^{-4} \text{ g}) (6.02 \times 10^{23} \text{ atoms}/M)}{14 \text{ g}/M}$$

$$= 2.95 \times 10^{19}$$

The half-life of ${}^{14}C$ is 5580 years. Hence, the activity of ${}^{14}C$ is

$$A = \lambda N = \frac{0.693 \, N}{T_{1/2}}$$

$$= \frac{0 \cdot 693}{(5580 \text{ yr})(365 \text{ day/yr})(24 \text{ hr/day})(3600 \text{ sec/hr})} (2.95 \times 10^{19} \text{ atoms})$$

$$A = 1.16 \times 10^8 \text{ dps}$$

Since 1 mCi $= 3.7 \times 10^7$ dps

$$A = \frac{1.16 \times 10^8 \text{ dps}}{3.7 \times 10^7 \text{ dps/mCi}}$$

$$= 3.13 \text{ mCi}$$

Example 2-4

What activity of ${}^{32}P$ as Na_3PO_4 must be ordered at 12:00 PM Friday to have at least 50 mCi at 8:00 AM on the following Wednesday? The time between ordering and use is 4.83 days.

$$A = A_0 \, e^{-\lambda t} = A_0 \, e^{-(0.693/T_{1/2})t}$$

$$50 \text{ mCi} = A_0 \, e^{-(0.693/14.3 \text{days})(4.83\text{days})}$$

$$= A_0 \, e^{-0.234}$$

$$A_0 = (50 \text{ mCi})/(e^{-0.234})$$

$$= (50 \text{ mCi}) (1.265)$$

$$A_0 = 63 \text{ mCi}$$

Transient and Secular Equilibrium

If a daughter nuclide formed by decay of a parent also is radioactive, and if the half-life of the daughter is shorter than the half-life of the parent, then a condition of transient equilibrium between parent and daughter may be established. Before transient equilibrium is established, the rate of production of the daughter exceeds its rate of decay, and the daughter activity increases with time. As the daughter nuclide attains its greatest activity, the rate of decay reaches the rate of production, and the activity of the daughter momentarily remains constant. Beyond this moment of equilibrium between production and decay, the rate of decay of the daughter exceeds its rate of production, and the daughter decays with an "apparent" half-life equal to the longer half-life of the parent.

The number N_2 of daughter atoms present at any time t is

$$N_2 = (N_2)_0 + \frac{\lambda_1}{\lambda_2 - \lambda_1}(N_1)_0(e^{-\lambda_1 t} - e^{-\lambda_2 t})$$

where $(N_1)_0$ and $(N_2)_0$ are the number of parent and daughter atoms, respectively, present when $t = 0$, and λ_1 and λ_2 are the respective decay constants of the parent and daughter. After the daughter has attained its maximum activity and is decaying with an apparent half-life equal to the half-life of the parent, the ratio of activities of parent and daughter is

$$\frac{A_1}{A_2} = \frac{\lambda_2 - \lambda_1}{\lambda_2}$$

Hence, the activity of the daughter is greater than the activity of the parent for two nuclides in transient equilibrium.

Transient equilibrium for the transition

$$^{132}\text{Te}\ (T_{1/2} = 78\ \text{hr}) \rightarrow\ ^{132}\text{I}\ (T_{1/2} = 2.3\ \text{hr})$$

is illustrated in Figure 2-13. Initially, the activity of the daughter ^{132}I increases rapidly, because few daughter atoms have been formed and the number which decay is relatively small. The ^{132}I activity is greatest when parent and daughter activities are equal. At all later times, the activity of

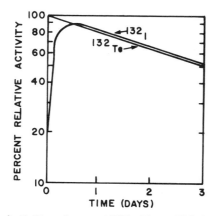

Figure 2-13 Activities of parent ^{132}Te ($T_{1/2} = 78$ hr) and daughter ^{132}I ($T_{1/2} = 2.3$ hr) as a function of time, illustrating the establishment of transient equilibrium between two nuclides when the half-life of the parent is not much greater than the half-life of the daughter.

the daughter exceeds that of the parent, and both isotopes appear to decay with the half-life of the parent.

Example 2-5

For the transition $^{132}\text{Te} \rightarrow {}^{132}\text{I}$, determine the activity of ^{132}I 6 hr and 24 hr after isolation of 100 mCi of pure ^{132}Te.

After 6 hr

$$N_2 = (N_2)_0 + \frac{\lambda_1}{\lambda_2 - \lambda_1} (N_1)_0 (e^{-\lambda_1 t} - e^{-\lambda_2 t})$$

$$(N_2)_0 = 0$$

$$\lambda_1 = \frac{0.693}{(T_{1/2})_1} \qquad \lambda_2 = \frac{0.693}{(T_{1/2})_2}$$

$$= \frac{0.693}{78 \text{ hr}} \qquad = \frac{0.693}{2.3 \text{ hr}}$$

$$= 0.0089/\text{hr} \qquad = 0.30/\text{hr}$$

$$(N_1)_0 = \frac{(A_1)_0}{\lambda_1} = \frac{(100 \text{ mCi})(3.7 \times 10^7 \text{ dps/mCi})}{0.693/[(78 \text{ hr})(3600 \text{ sec/hr})]}$$

$$= 1.5 \times 10^{15} \text{ atoms}$$

$$N_2 = \frac{0.0089/\text{hr}}{(0.30 - 0.0089)/\text{hr}} (1.5 \times 10^{15} \text{ atoms}) (e^{-(0.0089/\text{hr})(6\text{hr})} - e^{(0.30/\text{hr})(6\text{hr})})$$

$$= (4.57 \times 10^{13} \text{ atoms}) (0.948 - 0.164)$$

$$= 3.58 \times 10^{13} \text{ atoms}$$

$$A = \lambda N$$

$$= \frac{0.693}{T_{1/2}} N$$

$$= \frac{0.693}{(2.3 \text{ hr})(3600 \text{ sec/hr})} (3.58 \times 10^{13} \text{ atoms})$$

$$= 3.0 \times 10^9 \text{ dps}$$

$$= \frac{3.0 \times 10^9 \text{ dps}}{3.7 \times 10^7 \text{ dps/mCi}}$$

$$A = 81 \text{ mCi}$$

After 24 hr, the ratio of ^{132}Te to ^{132}I activity is

$$\frac{A_1}{A_2} = \frac{\lambda_2 - \lambda_1}{\lambda_2}$$

$$= \frac{(0.30 - 0.0089)/\text{hr}}{0.30/\text{hr}}$$

$$= 0.97$$

After 24 hr, the activity of ^{132}Te is

$$A_1 = A_0 e^{-\lambda t}$$

$$= (100 \text{ mCi}) e^{-(0.693/78\text{hr})(24\text{hr})}$$

$$= (100 \text{ mCi}) e^{-0.213}$$

$$= (100 \text{ mCi}) (0.808)$$

$$A_1 = 80.8 \text{ mCi}$$

The activity of ^{132}I is

$$A_2 = \frac{A_1}{0.97}$$

$$= \frac{80.8 \text{ mCi}}{0.97}$$

$$A_2 = 83.3 \text{ mCi}$$

If the half-life of the parent is much greater than the half-life of the daughter, then the activity of the daughter remains essentially constant after equilibrium between production and decay of the daughter has been achieved. Under these circumstances, the equation for the number N_2 of daughter atoms may be simplified to

$$N_2 = (N_2)_0 + \frac{\lambda_1}{\lambda_2} (N_1)_0 (1 - e^{-\lambda_2 t})$$

After several half-lives of the daughter have elapsed, $e^{-\lambda_2 t} \simeq 0$, and

$$N_2 = (N_2)_0 + \frac{\lambda_1}{\lambda_2} (N_1)_0$$

If $(N_2)_0 = 0$, then

$$\frac{(N_1)_0}{N_2} = \frac{\lambda_2}{\lambda_1}$$

Since the parent activity $A_1 = \lambda_1 N_1$ (where $\lambda_1 N_1 = \lambda_1 (N_1)_0$, because the number of parent atoms remains essentially constant) and the daughter activity $A_2 = \lambda_2 N_2$, the activities of parent and daughter are equal after establishment of secular equilibrium (Figure 2-14).

Example 2-6

Compute the activity and number of atoms of ^{222}Rn in secular equilibrium with 1 mCi of ^{226}Ra.

Since the activities of parent and daughter are equal when two

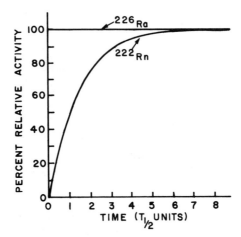

Figure 2-14 Growth of activity and secular equilibrium of ^{222}Rn ($T_{1/2}$ = 3.83 days) produced by decay of ^{226}Ra ($T_{1/2}$ = 1600 years).

nuclides are in secular equilibrium, the activity of ^{222}Rn is also 1 mCi. The number N_2 of atoms of ^{222}Rn is

$$A_2 = \lambda_2 N_2$$

$$N_2 = \frac{A_2}{\lambda_2}$$

$$= \frac{A_2 (T_{1/2})}{0.693}$$

$$= \frac{(1 \text{ mCi})[3.7 \times 10^7 \text{ dps/mCi } (3.83 \text{ days})(24 \text{ hr/day})(3600 \text{ sec/hr})]}{0.693}$$

$$N_2 = 1.77 \times 10^{13} \text{ atoms}$$

Data for Radioactive Nuclides

Decay schemes for many radioactive nuclei are listed in the *Radiological Health Handbook*[1], *Table of Isotopes*[2], Supplements 2 and 4 of the *Journal of Nuclear Medicine*[3], *Radioatoms in Nuclear Medicine*[4], and *Nuclear Data Sheets*[5]. Charts of the nuclides, available from the General Electric Company in Schenectady, N.Y., the United States Atomic Energy Commission in Washington, D.C., and the Mallinckrodt Chemical Works in St. Louis, Mo., contain useful data concerning the decay of radioactive nuclides. A section from a typical chart is reproduced in Figure 2-15. In this section, isobars are positioned along 45° diagonals, isotones along vertical lines, and isotopes along horizontal lines.

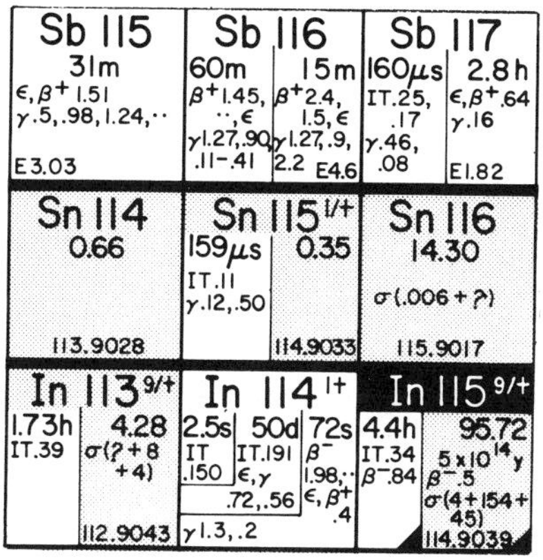

Sb 115	Sb 116		Sb 117	
31m	60m	15m	160μs	2.8h
ϵ,β^+1.51	β^+1.45,	β^+2.4,	IT.25,	ϵ,β^+.64
γ.5,.98,1.24,··	··,ε	1.5,ε	.17	γ.16
	γ1.27,90	γ1.27,.9,	γ.46,	
	.11–.41	2.2 E4.6	.08	
E 3.03			E1.82	

Sn 114	Sn 115 $^{1/+}$		Sn 116	
0.66	159μs	0.35	14.30	
	IT.11			
	γ.12,.50		σ(.006 + ?)	
113.9028	114.9033		115.9017	

In 113 $^{9/+}$	In 114 $^{1+}$			In 115 $^{9/+}$	
1.73h 4.28	2.5s	50d	72s	95.72	
IT.39 σ(?+8	IT	IT.191	β^-	4.4h	
+4)	.150	ε,γ	1.98,·	IT.34	5×10^{14} y
		.72,.56	ε,β^+	β^-.84	β^-.5
			.4		σ(4+154+ 45)
112.9043	γ1.3,.2			114.9039	

Figure 2-15 Section from a chart of the nuclides. (Courtesy of Knolls Atomic Power Laboratory, Schenectady, N.Y., operated by the General Electric Company for the United States Atomic Energy Commission.)

The energy of radiations emitted by various nuclei is expressed in units of million electron volts.

PROBLEMS

1. The half-life of ^{99m}Tc is 6.0 hr. What time is required for 200 mCi of ^{99m}Tc to decay to 50 mCi? What time is required for the number of ^{99m}Tc atoms to equal $\frac{1}{8}$ of the original number? What time is required for 200 mCi of ^{99m}Tc to decay to 50 mCi, if the ^{99m}Tc is in transient equilibrium with ^{99}Mo ($T_{1/2} = 67$ hr)?

2. What is the mass in grams of a water sample containing 50 mCi of tritiated water (H^3HO), if the sample is diluted 10^6: 1 by volume with ordinary water? How many tritium atoms are in 50 mCi of H^3HO?

3. Some ^{210}Bi nuclei decay by α emission and others by negatron emission. Write the equation for each mode of decay and identify the daughter nuclide.

4. What are the frequency and wavelength of a 250 keV γ-ray?

5. ^{108}Ag nuclei may decay by negatron emission, positron emission, or electron capture. Write the equation for each mode of decay and identify the daughter nuclide.

6. From a chart of the nuclides determine:

   (a) Whether ^{14}C is stable or unstable

   (b) Whether ^{11}C decays by negatron or positron emission

   (c) The nuclide which decays to ^{12}C by positron emission

   (d) The nuclide which decays to ^{12}C by negatron emission

   (e) The half-life of ^{10}C

   (f) The percent abundance of ^{12}C in naturally occurring carbon

   (g) The isotopic mass of ^{13}C

   (h) The energy of γ-rays emitted by ^{15}C

   (i) The maximum energy of positrons from ^{10}C

7. How many atoms and grams of ^{90}Y ($T_{1/2} = 64.2$ hr) are in secular equilibrium with 50 mCi of ^{90}Sr ($T_{1/2} = 28.8$ years)?

8. The only stable isotope of fluorine is ^{19}F. What modes of decay would be expected for ^{18}F and ^{20}F?

9. For a nuclide X with the decay scheme

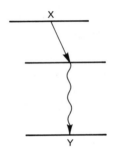

how many γ-rays are emitted per 100 disintegrations of X, if the coefficient of internal conversion is 0.33?

10. How many millicuries of ^{47}Ca ($T_{1/2} = 4.5$ days) must be shipped 8:00 AM on Monday to have 50 mCi on noon of the next Friday?

REFERENCES

[1] U.S. Department of Health, Education, and Welfare, Office of Technical Services. 1970. Radiological health handbook. Revised.

[2] Lederer, C., J. Hollander, and I. Perlman. 1967. Table of isotopes. John Wiley & Sons, New York.

[3] Dillman, L. 1969–1970. Radionuclide decay schemes and nuclear parameters

for use in radiation-close estimation. J. Nucl. Med. (Parts 1 and 2, Suppl. 2 and 4).

[4] Blichert-Toft, P. 1968. Radioatoms in nuclear medicine. Rigmor Nilsson, Gothenburg, Sweden.

[5] Way, K. (ed.). Nuclear data sheets. Academic Press, New York.

3

Interactions of Particulate Radiation

If the sum of the kinetic energies of two particles after they have interacted equals the sum of the kinetic energies of the particles before interaction, then the interaction is termed an *elastic interaction*. If the sum of the kinetic energies is changed by the interaction, then the interaction is inelastic. Heavy charged particles such as protons, deuterons, and α particles usually interact inelastically with electrons of atoms along the path of the particles. Lighter charged particles such as negatrons and positrons interact with both the electrons and the nuclei of atoms in the absorbing medium. Neutrons are uncharged particles and have no associated electric field to interact with the electric fields of electrons and nuclei in the medium. These particles interact primarily by elastic and inelastic collisions with absorber nuclei.

Interactions of Heavy Charged Particles

Heavy charged particles such as protons, deuterons, and α particles lose energy rapidly as they penetrate matter. Most of the energy is lost during inelastic interactions with the electrons of the absorbing medium. The transfer of energy is accomplished by the interaction of the electric field of the particle with the electric fields of electrons in the medium, and physical contact between the interacting particles is not required. Usually, the transfer of energy results in the displacement of electrons in absorber atoms to orbits farther from the nucleus. This process is termed *excitation*. If the amount of energy transferred to an electron is sufficient to

34

eject the electron from the atom, the process is termed *ionization*. The ejected electron and residual positive ion are referred to collectively as an *ion pair* (IP). Often the ejected electron has enough kinetic energy to produce a few additional ion pairs in the immediate vicinity of the original positive ion. The resulting concentration of ion pairs is termed an *ion cluster*.

The average energy expended by a charged particle per ion pair produced is termed the *W* quantity for the particle in the particular absorbing medium. The *W* quantity for charged particles in air is 33.7 eV. The energy required to remove an electron from nitrogen or oxygen, the major constituents of air, is much less than 33.7 eV. The *W* quantity includes not only energy expended in ionizing atoms, but also energy lost as the charged particle excites atoms, interacts with nuclei, and increases the rate of vibration of nearby molecules. On the average, 2.2 atoms are excited for each ion pair produced by charged particles in air.

The specific ionization (SI) is the number of ion pairs produced per unit length of path of a charged particle. The SI of α particles in air varies from about 30,000 IP/cm over most of their path length to perhaps 70,000 IP/cm near the end of their path, where the particles are moving slowly. The SI of heavy particles (for example, $^2H^+$, $^1H^+$) with half the charge of an α particle is considerably less than that for α particles of the same energy.

The linear energy transfer (LET) is the average energy deposited per unit length of path of a charged particle. For heavy charged particles, the LET is approximately the product of the SI and the *W* quantity.

$$LET = (SI)(W)$$

Example 3-1
If the SI is 30,000 IP/cm at a particular location along the path of an α particle in air, what is the approximate LET at the location?

$$
\begin{aligned}
LET &= (SI)(W) \\
&= (30{,}000 \text{ IP/cm})(33.7 \text{ eV/IP}) \\
&= 1 \text{ MeV/cm}
\end{aligned}
$$

The range of charged particles in a particular medium is the straight-line distance traversed by the particles before they are stopped completely. For heavy particles with energy *E*, the range in a particular medium may be estimated from the LET averaged over the entire range of the particles.

$$range = \frac{E}{(LET)_{avg}}$$

Example 3-2

What is the range in air for 4 MeV α particles with a $(LET)_{avg}$ of 1 MeV/cm?

$$\text{range} = \frac{E}{(LET)_{avg}}$$
$$= \frac{4\,\text{MeV}}{1\,\text{MeV/cm}}$$
$$= 4.0\,\text{cm}$$

From Examples 3-1 and 3-2, it is apparent that α particles (and other heavy charged particles) produce dense ionization but have a limited range. Since the density of soft tissue (1 g/cm³) is about 1000 times greater than the density of air (0.00129 g/cm³), heavy charged particles penetrate soft tissue to a depth of only a few microns (1 μ = 10^{-4} cm). For example, α particles of a few million electron volts from a radioactive source near or on the body penetrate only the most superficial layers of skin, and are not hazardous so long as they are emitted by a source outside the body. Because of the dense ionization produced by these particles, however, α-emitting sources may be very hazardous if present within the body. Beams of heavy charged particles from an accelerator also may be hazardous, because the energy of these particles may be much greater than the energy of α particles emitted by radioactive atoms.

The SI and LET are not constant along the entire path of a heavy charged particle traversing a homogeneous medium. As the velocity of the particle decreases rapidly near the end of its path, nearby electrons are subjected to the electric field of the particle for a longer period of time, and the probability of excitation and ionization is increased. Consequently, the SI and LET increase near the end of the particle's path,

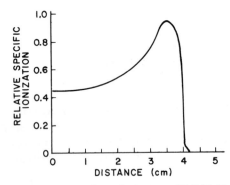

Figure 3-1   Specific ionization of a beam of 5.3 MeV α particles from [210]Po as a function of distance traversed in air.

and a peak (the "Bragg peak") is apparent in a plot of SI (or LET) as a function of distance traversed by the particle (Figure 3-1).

Interactions of Electrons

Negative and positive electrons (that is, negatrons and positrons) traversing a medium may interact with both electrons and nuclei in the medium. During these interactions, the impinging electrons lose energy and are deflected or scattered from their original direction of motion. Interactions with electrons of the medium produce excitation and ionization, with the kinetic energy E_k of an ejected electron equal to the energy E received from the impinging electron minus the ejected electron's binding energy E_B.

$$E_k = E - E_B.$$

If the electron is ejected from an orbit where the binding energy is negligible compared to the energy received, then the interaction may be considered an elastic collision between "free" particles. If the binding energy is not negligible, then the interaction must be considered inelastic.

Impinging electrons are scattered with a probability which increases with the electron density (electrons per gram) of the absorbing medium and decreases with the energy of the impinging electrons. The SI of negatrons and positrons in air ranges from 50 IP/cm to 500 IP/cm; this SI is about 1/100 that for α particles in the same medium.

After expending its kinetic energy, a positron combines with a negative electron in the absorbing medium. The two particles annihilate each other and their mass appears as electromagnetic radiation, usually two 0.51 MeV photons moving in opposite directions. This interaction is termed *pair annihilation*, and the photons are referred to as *annihilation photons* or *annihilation radiation*.

Electrons also may interact elastically and inelastically with nuclei of an absorbing medium. The frequency of elastic interactions with nuclei increases rapidly with the atomic number of the nuclei and decreases with the kinetic energy of the electrons. Backscattering of negatrons and positrons (Chapter 19) is due primarily to elastic scattering of these particles by nuclei.

The probability for elastic scattering of electrons by electrons and by nuclei of an absorber is about equal if the absorber is hydrogen ($Z = 1$). In absorbers of higher Z, elastic scattering by nuclei occurs more frequently, because the probability for scattering by nuclei increases with Z^2, whereas the probability for scattering by electrons varies with Z.

Negatrons and positrons also may be scattered inelastically during interactions with absorber nuclei. During these interactions, energy is released in the form of electromagnetic radiation, termed *bremsstrahlung* (braking radiation). A bremsstrahlung photon may possess any energy up to the entire kinetic energy of the impinging electron (Figure 3-2). Bremsstrahlung photons are released approximately at right angles to the motion of low-energy electrons; as the energy of the electrons increases, the angle is reduced. The rate of energy loss by bremsstrahlung production increases with Z^2 of the absorbing medium and linearly with the energy of the electrons. The shape of the energy spectrum for bremsstrahlung photons is independent of the atomic number of the absorber.

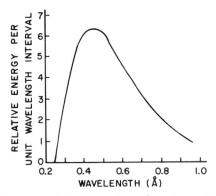

Figure 3-2   Bremsstrahlung spectrum for a tungsten target bombarded by 50 keV electrons[1].

For electrons, the ratio of energy lost by "radiative" interactions (bremsstrahlung production) to energy lost by "collisional" interactions (excitation and ionization) is approximately

$$\frac{E_{\text{rad}}}{E_{\text{coll}}} = \frac{EZ}{820}$$

where E is the energy of the impinging electrons in million electron volts and Z is the atomic number of the absorbing medium. For example, radiative and collisional energy losses are about equal for 10 MeV electrons traversing lead ($Z = 82$).

When charged particles move through a medium at a velocity greater than the velocity of light in the medium, energy is radiated by the particles in the form of blue light. The visible light is termed *Cerenkov radiation*, and is responsible for a small proportion of the total energy lost by electrons moving at high speeds in the medium. The blue light emanating from the vicinity of the core of a "swimming pool reactor" is Cerenkov radiation.

Measurement of E_{max} for β Particles

Energy spectra for negatrons and positrons may be determined accurately with a β-ray spectrometer. From these spectra, the maximum energy may be determined for negatrons or positrons emitted by a radioactive sample. A simpler but less accurate method for determining E_{max} has been developed from a procedure developed originally by Feather[2]. With this method, the number of negatrons or positrons transmitted by a foil of absorbing material is measured as the thickness of the foil is varied. Typical data, corrected for background and γ emission from the sample, are plotted in Figure 3-3.

The half-value thickness $D_{1/2}$ is the thickness of absorber which reduces the number of particles reaching the detector to half. A value for $D_{1/2}$ may be determined from the straight-line portion of the transmission

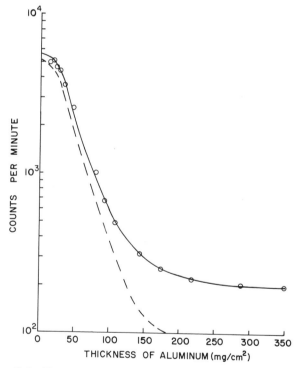

Figure 3-3 The transmission of negatrons from ^{134}Cs through aluminum. The tail in the curve at large thicknesses of absorber reflects the emission of γ-rays from the ^{134}Cs source and the production of bremsstrahlung in the absorber. The dashed curve represents transmission data corrected for the presence of γ-rays.

curve. With a value of $D_{1/2}$ in units of milligrams per square centimeter,[*] the maximum energy E_{max} in million electron volts may be computed for negatrons or positrons emitted by the sample:

$$D_{1/2} = 41E_{max}^{1.14}$$

This equation is useful for negatrons and positrons with a maximum energy between 0.1 MeV and 3 MeV.

The modified Feather method provides values for E_{max} which are accurate to within $\pm5\%$. Measurement of E_{max} often helps to identify nuclides present in a counting sample.

Example 3-3

From a transmission curve for negatrons from [204]Tl, the half-value thickness in aluminum is 30.4 mg/cm². Compute the maximum energy of negatrons emitted by this nuclide.

$$D_{1/2} = 41E_{max}^{1.14}$$

$$\frac{D_{1/2}}{41} = E_{max}^{1.14}$$

$$\log\frac{D_{1/2}}{41} = \log E_{max}^{1.14}$$

$$= 1.14\log E_{max}$$

$$\log E_{max} = \frac{1}{1.14}\log\frac{D_{1/2}}{41}$$

$$= \frac{1}{1.14}\log\frac{30.4}{41}$$

$$= \frac{1}{1.14}\log 0.74$$

$$= \left(\frac{1}{1.14}\right)(-0.13)$$

$$= -0.114$$

$$E_{max} = 0.77\ \text{MeV}$$

The range is the thickness of a particular absorber required to absorb the most energetic negatron or positron from a radioactive sample. From Figure 3-4, the maximum energy may be determined if the range of the particles in aluminum is known. The range of negatrons or positrons from a radioactive nuclide may be estimated roughly from a transmission curve

---

[*]The thickness of an absorber in units of milligrams per square centimeter is computed by:
(1) thickness in mg/cm² = (thickness of absorber in cm)(density of absorber in mg/cm³);
(2) thickness in mg/cm² = (mass of absorber in mg)/(cross-sectional area of absorber in cm²).
Half-value thicknesses in units of milligrams per square centimeter are nearly independent of the density and atomic number of the absorber.

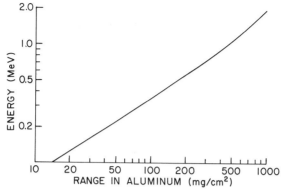

Figure 3-4   Range-energy curve for β particles in aluminum.

similar to Figure 3-3. However, this approach is inaccurate because:

(1) The ordinate (*y* axis) is logarithmic and a transmission of zero cannot be plotted

(2) With large thicknesses of absorber, the limited number of particles incident upon the detector increases the influence of statistical fluctuations on the measured count rate

(3) The "straggling" of negatrons or positrons at large thicknesses of absorber, together with the presence of γ-rays and bremsstrahlung, create a "tail" in the transmission curve; it is difficult to correct the shape of the curve for these effects

Interactions of Neutrons

Nuclear reactors (Chapter 5) are the source of neutrons used most widely for the irradiation of chemical and biological specimens, for the production of radioactive isotopes, and for other applications in biological and chemical research. However, neutron beams also are available from neutron generators and cyclotrons, in which low-*Z* nuclei (for example, ^{3}H or ^{9}Be) are bombarded by high-velocity, positively charged particles (for example, ^{1}H$^+$, ^{2}H$^+$, or ^{4}He^{2+}). For example, 14.1 MeV neutrons may be produced in a neutron generator by bombarding a tritium target with deuterons (^{2}H$^+$) accelerated through a potential difference of 150 kV. The energy distribution of neutrons from neutron generators and cyclotrons depends upon the target material and upon the type and energy of the bombarding particle.

Another neutron source is ^{252}Cf, an artificially produced nuclide which fissions spontaneously with a half-life of 2.65 years. Also, low intensity sources of neutrons may be fabricated by intimate mixing of a source of

α particles with a suitable target material, usually a low-Z element. Typical sources are ^{226}Ra:Be, ^{210}Po:Be, and ^{239}Pu:Be. The Ra:Be source has a long half-life, but offers a neutron flux contaminated severely with γ-rays. The Po:Be source contributes few γ-rays, but has a half-life of only 138 days. The Pu:Be source has a long half-life and furnishes few γ-rays, but provides fewer neutrons than the other two sources.

Neutrons are uncharged particles which, at higher velocities, usually interact by "billiard ball" or "knock-on" elastic collisions with absorber nuclei. The energy transfer from neutron to nucleus is greatest if the mass of the nucleus is about equal to that of the neutron. Consequently, fast neutrons are slowed or "thermalized" most rapidly by interactions with hydrogen nuclei, and materials with a high content of hydrogen (for example, H_2O and paraffin) are used often as shielding for beams of fast neutrons. Other low-Z materials such as heavy water (water enriched with deuterium) and graphite are used to thermalize fast neutrons in nuclear reactors (Chapter 5). In soft tissue, elastic collisions with hydrogen nuclei account for 90% of the energy degradation of a beam of neutrons with energies less than 10 MeV.

Fast neutrons also may be captured by a nucleus, causing emission from the nucleus of one or more neutrons, a γ-ray, or a charged particle such as a proton, deuteron, or α particle. Capture of fast neutrons by nuclei accounts for about 30% of the energy deposited in soft tissue by 14.1 MeV neutrons. A fast neutron also may initiate fission of some nuclei. These processes are described in greater detail in Chapter 5.

Slowly moving neutrons interact primarily by radiative capture, in which a nucleus captures the neutron and releases the excess energy as one or more γ rays. For many nuclei, the probability of radiative capture varies as $1/v$, where v is the velocity of the neutron. When the energy of impinging neutrons is matched closely to a difference in the energy levels of a nucleus, resonance absorption of slow neutrons also occurs.

PROBLEMS

1. Electrons with kinetic energy of 100 keV have a SI of about 130 IP/cm in air. Assuming that elastic and inelastic interactions with nuclei provide a negligible contribution to the LET of these electrons, estimate the LET of 100 keV electrons in air.

2. Estimate the ratio of radiative energy loss to collisional energy loss for 100 keV electrons traversing air (effective atomic number = 7.65).

3. What is the thickness in milligrams per square centimeter for a sheet of

aluminum 0.50 mm thick (density $Al = 2.7$ g/cm^3)? For a layer of air 5.0 cm thick (density air $= 1.29 \times 10^{-3}$ g/cm^3)?

4. The measured half-value thickness $D_{1/2}$ for β particles from ^{134}Cs is 25.0 mg/cm^2. What is the maximum β energy estimated by the modified Feather method?

5. From Figure 3-4, what is the range in aluminum for β particles from ^{90}Sr ($E_{max} = 0.55$ MeV)?

REFERENCES

[1] Ulrey, C. 1918. An experimental investigation of the energy in the continuous x-ray spectra of certain elements. Phys. Rev. 11:401.

[2] Feather, N. 1938. Further possibilities for the absorption method of investigating the primary β particles from radioactive substances. Proc. Cambridge Phil. Soc. 34:599.

4

Interactions of
X Radiation
and γ Radiation

When charged particles interact with electrons and nuclei of an absorbing medium, x-rays are produced; γ-rays are released during the decay of radioactive nuclei. Except for the difference in origin, x-rays and γ-rays are indistinguishable, and their types of interactions are identical. The interactions of greatest importance are photoelectric absorption, Compton scattering, and pair production. Coherent scattering and photodisintegration reactions are interactions of lesser significance to biologic uses of x- and γ-rays.

Photoelectric Absorption

During photoelectric interaction of an x- or γ-ray photon, the total energy of the photon is transferred to an electron of an atom in the absorbing medium (Figure 4-1). Usually, the electron occupies an inner electron orbit in the atom. The photon disappears, and the electron is ejected from the atom with a kinetic energy E_k:

$$E_k = h\upsilon - E_B$$

where $h\upsilon$ is the energy of the photon and E_B is the binding energy of the electron. If the photon energy is relatively low, the ejected electron, termed a *photoelectron*, is released approximately at 90° to the motion of the photon. For photons of higher energy, the direction of the photoelectron is closer to that of the incident photon.

44

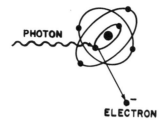

PHOTON

ELECTRON

Figure 4-1 During a photoelectric interaction, the entire energy $h\nu$ of an x- or γ-ray photon is transferred to an electron of an atom in the medium. The photon disappears and the electron is ejected from the atom with a kinetic energy $E_k = h\nu - E_B$, where E_B is the binding energy of the electron.

Example 4-1

What is the kinetic energy of a photoelectron ejected from the K shell of cadmium ($E_B = 26.7$ keV) by photoelectric interaction of an 80.0 keV γ-ray?

$$E_k = h\nu - E_B$$
$$= 80.0 \text{ keV} - 26.7 \text{ keV}$$
$$= 53.3 \text{ keV}$$

The ejection of an electron from an inner orbit of an atom creates a vacancy in the orbit which is filled immediately by an electron from a location farther from the nucleus. As the atom regains structural stability by the cascading of electrons among orbits, characteristic x-rays and Auger electrons are released. The Auger electrons and, often, the x-rays are absorbed rapidly in the surrounding medium.

The probability of photoelectric absorption depends strongly upon the atomic number of the absorbing medium. In general, the probability of photoelectric absorption varies with Z^3 of the medium. For example, the probability of photoelectric absorption in aluminum ($Z = 13$) is about eight times greater than the probability of photoelectric interaction in an equal mass of carbon ($Z = 6$). The strong influence of atomic number on the probability of photoelectric absorption is the principal reason why high-Z materials such as lead are used to shield radiation detectors and γ-ray sources.

In biologic tissue, photoelectric absorption is the dominant interaction for x- and γ-ray photons below about 35 keV. As the energy of the photons increases, the probability of photoelectric interaction decreases rapidly (approximately as $(1/(h\nu)^3)$). This rapid decrease is described in Figure 4-2, where the mass attenuation coefficient (in units of square centimeters per gram) for photoelectric absorption is plotted as a function

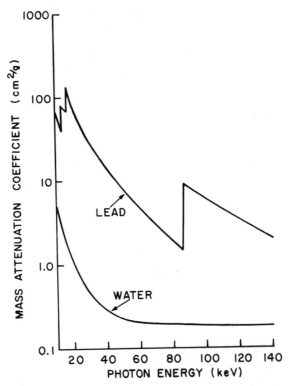

Figure 4-2 Mass attenuation coefficient for photoelectric absorption of x- and γ-ray photons in lead and water, plotted as a function of the photon energy in keV. K- and L-absorption edges are depicted for lead.

of the energy of photons traversing lead and water. Discontinuities (absorption edges) in the curve for lead occur at photon energies equal to the binding energy of electrons in the K and L orbits of lead. Photons with energy less than the binding energy of electrons in a particular orbit do not interact photoelectrically with those electrons. For example, photons with energy less than the binding energy (88 keV) of K electrons in lead interact only with electrons in orbits farther from the nucleus. The probability of interaction with electrons in a particular orbit is greatest when the photon energy is just equal to or slightly greater than the binding energy of the electrons. Hence, the probability of photoelectric interaction increases rapidly for photons with energy equal to the binding energy of electrons in a particular orbit, and discontinuities or absorption edges are created in the curve of mass attenuation coefficient versus

energy. Absorption edges for photoelectric absorption in water occur at very low energies and are not shown in Figure 4-2.

Compton Scattering

During a Compton interaction, part of the energy of an incident x- or γ-ray photon is transferred to an electron, usually a loosely bound or "free" electron, in the absorbing medium. The electron recoils at an angle θ with respect to the motion of the incident photon, and the photon is scattered with reduced energy at an angle ϕ (Figure 4-3). The binding energy of the electron usually is negligible, and the kinetic energy E_k of the Compton electron equals the energy lost by the photon. That is,

$$E_k = h\upsilon - h\upsilon'$$

where $h\upsilon$ is the photon energy before interaction, and $h\upsilon'$ is the photon energy after interaction. Although the photon may be scattered at any angle ϕ, the direction of the Compton electron is confined to an angle θ less than 90° with respect to the direction of the incident photon. The angles ϕ and θ tend to decrease as the energy of the incident photon increases.

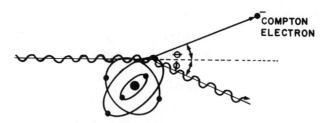

Figure 4-3   During a Compton interaction, part of the energy of an incident photon is transferred to a loosely bound electron in the absorbing medium. The electron is ejected at an angle θ, and the photon is scattered with reduced energy at an angle ϕ.

The change in wavelength $\Delta\lambda$ of a photon undergoing Compton scattering is:

$$\Delta\lambda = 0.0243(1 - \cos\phi)$$

where ϕ is the angle of scatter of the photon. The wavelength λ', of the scattered photon is

$$\lambda' = \lambda + \Delta\lambda$$

where λ is the wavelength of the incident photon. The energies $h\upsilon$ and

hv' of the incident and scattered photons are

$$hv = \frac{12.4}{\lambda}$$

$$hv' = \frac{12.4}{\lambda'}$$

with the energy in kiloelectron volts and the wavelength in angstroms.

Example 4-2

A 300 keV photon is scattered at a 45° angle during a Compton interaction. What are the wavelength λ' and the energy hv' of the scattered photon?

The wavelength λ of the incident photon is

$$hv = \frac{12.4}{\lambda}$$

$$\lambda = \frac{12.4}{hv}$$

$$= \frac{12.4}{300 \text{ keV}}$$

$$= 0.041 \text{ Å}$$

The change in wavelength $\Delta\lambda$ during the interaction is

$$\Delta\lambda = 0.0243(1 - \cos \phi)$$
$$= 0.0243(1 - \cos 45°)$$
$$= 0.0243(1 - 0.71)$$
$$= 0.007 \text{ Å}$$

The wavelength λ' of the scattered photon is

$$\lambda' = \lambda + \Delta\lambda$$
$$= 0.041 \text{ Å} + 0.007 \text{ Å}$$
$$= 0.048 \text{ Å}$$

The energy hv' of the scattered photon is

$$hv' = \frac{12.4}{\lambda'}$$

$$= \frac{12.4}{0.048 \text{ Å}}$$

$$= 258 \text{ keV}$$

If the binding energy of the Compton electron is negligible, then the kinetic energy E_k of the Compton electron is

$$E_k = hv - hv'$$
$$= 300 \text{ keV} - 258 \text{ keV}$$
$$= 42 \text{ keV}$$

Example 4-3

Compute the maximum energy loss during Compton scattering of (a) a 25 keV γ-ray photon and (b) a 2.5 MeV γ-ray photon.

A photon loses the greatest energy when it is scattered at an angle of 180°, because $\Delta\lambda$ is greatest when $\phi = 180°$. Consequently, the energy lost as photons are scattered at an angle of 180° will be computed in this example.

For both 25 keV and 2.5 MeV photons scattered at 180°, the change in wavelength $\Delta\lambda$ is

$$
\begin{aligned}
\Delta\lambda &= 0.0243(1 - \cos \phi) \\
&= 0.0243(1 - \cos 180°) \\
&= 0.0243(1 - (-1)) \\
&= 0.049 \text{ Å}
\end{aligned}
$$

For the 25 keV photons, the wavelength λ is

$$
\begin{aligned}
\lambda &= \frac{12.4}{h\upsilon} \\
&= \frac{12.4}{25.0 \text{ keV}} \\
&= 0.490 \text{ Å}
\end{aligned}
$$

The wavelength λ' of a 25 keV photon scattered at 180° is

$$
\begin{aligned}
\lambda' &= \lambda + \Delta\lambda \\
&= 0.490 \text{ Å} + 0.049 \text{ Å} \\
&= 0.539 \text{ Å}
\end{aligned}
$$

The energy $h\upsilon'$ of the scattered photon is

$$
\begin{aligned}
h\upsilon' &= \frac{12.4}{\lambda'} \\
&= \frac{12.4}{0.539 \text{ Å}} \\
&= 23 \text{ keV}
\end{aligned}
$$

When a low-energy photon undergoes Compton scattering, most of the energy of the incident photon is retained by the scattered photon. Only a small fraction of the energy is transferred to the Compton electron.

For the 2.5 MeV photon, the wavelength λ is

$$
\begin{aligned}
\lambda &= \frac{12.4}{h\upsilon} \\
&= \frac{12.4}{2500 \text{ keV}} \\
&= 0.0049 \text{ Å}
\end{aligned}
$$

The wavelength λ' of a 2.5 MeV photon scattered at 180° is

$$\lambda' = \lambda + \Delta\lambda$$
$$= 0.0049 \text{ Å} + 0.049 \text{ Å}$$
$$= 0.054 \text{ Å}$$

The energy $h\upsilon'$ of the scattered photon is

$$h\upsilon' = \frac{12.4}{\lambda'}$$
$$= \frac{12.4}{0.054 \text{ Å}}$$
$$= 230 \text{ keV}$$

When a high-energy photon is scattered by the Compton process, most of the energy is transferred to the Compton electron. Only a small fraction of the energy is retained by the scattered photon.

Irrespective of the energy of the incident photon, a photon scattered at 180° can retain an energy not greater than 256 keV. Similarly, a photon scattered at 90° can possess no more than 511 keV[1]. The probability of Compton scattering per unit mass of absorbing medium is nearly independent of the atomic number of the medium, and varies only with the electron density (electrons per gram) of the medium.

Pair Production

While near a nucleus in the attenuating medium, an x- or γ-ray photon with an energy greater than 1.02 MeV may interact by pair production. The photon disappears and two electrons, one positive and one negative, appear in its place (Figure 4-4). The energy equivalent to the mass of the two particles is 1.02 MeV, and photons with energy less than this amount do not interact by pair production. Photon energy in excess of 1.02 MeV appears as kinetic energy of the negatron-positron pair.

$$h\upsilon(\text{MeV}) = 1.02 \text{ MeV} + (E_k)_{e^-} + (E_k)_{e^+}$$

During pair production, the small amount of energy transferred to the recoiling nucleus usually may be neglected. Positive electrons released during pair production produce annihilation radiation identical to that produced by positrons from radioactive nuclei.

Occasionally, pair production occurs in the field of an electron rather than a nucleus. This interaction is termed *triplet production*, because energy is transferred to the electron and it is ejected from the atom, resulting in the release of three ionizing particles (two negative electrons and one positron). The threshold energy for triplet production is 2.04

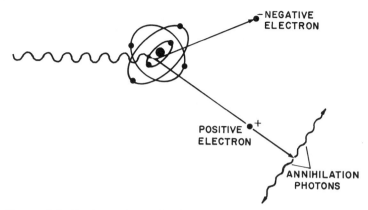

Figure 4-4 During pair production, the energy of a photon is transformed into the mass and kinetic energy of a negatron and positron.

MeV. The ratio of triplet to pair production increases with the energy of incident photons and decreases with increasing atomic number of the absorbing medium.

The mass attenuation coefficient for interaction of x- and γ-ray photons by pair production increases almost linearly with the atomic number of the medium and gradually with the energy of the incident photons. In water or soft tissue, only a small fraction of x- or γ-ray photons between 1.02 and 10 MeV interact by pair production.

Example 4-4

During a pair production interaction of a 4.00 MeV γ-ray photon, the residual energy is distributed equally between the negative and positive electron. What are the kinetic energies of these particles?

$$h\upsilon\,(\text{MeV}) = 1.02\ \text{MeV} + (E_k)_{e^-} + (E_k)_{e^+}$$

$$(E_k)_{e^-} = (E_k)_{e^+} = \frac{(h\upsilon - 1.02)\ \text{MeV}}{2}$$

$$= \frac{(4.00 - 1.02)\ \text{MeV}}{2}$$

$$(E_k)_{e^-} = (E_k)_{e^+} = 1.49\ \text{MeV}$$

The relative probability of photoelectric, Compton, and pair production interactions in media of different atomic numbers is plotted in Figure 4-5 as a function of photon energy. The curves in Figure 4-5 represent atomic numbers and photon energies for which adjacent effects are equally probable.

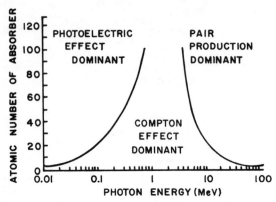

Figure 4-5   Relative importance of photoelectric, Compton, and pair production interactions as a function of atomic number of the absorbing medium and energy of the incident photons.

Coherent Scattering

In the interaction referred to as coherent scattering or Rayleigh scattering, an x- or γ-ray photon is scattered with negligible loss of energy during interaction with an electron. The interaction occurs usually with an electron in an inner orbit of an atom in the absorbing medium. Most often, the photon is scattered in a direction near that of the incident photon. In low-Z materials, coherent scattering is important only for photons with an energy less than 100 keV. The probability of coherent scattering increases with atomic number, and this interaction is important in high-Z materials for photon energies up to 1 MeV. Photons scattered coherently possess almost the same energy as the incident photons, and essentially no energy is transferred to atoms of the absorbing medium. Coherent scattering is responsible for interference patterns observed when low-energy x-rays impinge at selected angles upon a crystal.

Photodisintegration

A photodisintegration interaction between an x- or γ-ray photon and a nucleus is possible whenever the photon has enough energy to eject a neutron, a proton, or a particle of greater mass from the nucleus. To initiate a photodisintegration reaction, a photon usually must possess at least 8 MeV. However, photons of less energy sometimes undergo photo-disintegration interactions in some low-Z elements. Photodisintegration interactions with threshold energies of 1.65 MeV and 2.22 MeV are depicted below.

$$^9\text{Be}(h\upsilon, \text{n})^8\text{Be} \ (E_{\text{threshold}}) = 1.65 \ \text{MeV}$$

$$^2\text{H}(h\upsilon, \text{n})^1\text{H} \ (E_{\text{threshold}}) = 2.22 \ \text{MeV}$$

In the first reaction, ^9Be is the bombarded nucleus, $h\upsilon$ is the incident photon, n is the neutron ejected from the nucleus, and ^8Be is the nucleus remaining after the interaction.

X-Ray and γ-Ray Attenuation

The rate of removal of photons from an x- or γ-ray beam traversing a medium may be described by the expression

$$\text{rate of removal} = -\mu I$$

where I represents the number of photons in the beam and μ is the attenuation coefficient of the medium for the photons. The negative sign indicates that the number of photons in the beam decreases as the beam penetrates the medium. If all the photons in the beam possess the same energy (that is, the beam is monoenergetic), and if the photons are removed from the beam under conditions of good geometry (that is, the beam is narrow and contains no scattered photons), then the number of photons penetrating a thin slab of the medium of thickness x is

$$I = I_0 e^{-\mu x}$$

where I_0 represents the number of photons in the beam before the thin slab of medium is encountered. This equation is derived in Appendix 1. The number I_d of photons absorbed or scattered by the thin slab is

$$\begin{aligned} I_d &= I_0 - I \\ &= I_0 - I_0 e^{-\mu x} \\ &= I_0(1 - e^{-\mu x}) \end{aligned}$$

If the thickness x of the thin slab is expressed in units of length (for example, centimeters or inches), then the attenuation coefficient μ possesses units of 1/length (for example, 1/cm or 1/in.) and is termed a *linear attenuation coefficient*. If the thickness x is described in units of grams (or milligrams) per square centimeter, then the attenuation coefficient has units of square centimeters per gram (or per milligram), and is termed a *mass attenuation coefficient*. The mean free path or relaxation length is the average distance traversed by an x- or γ-ray photon before interaction in a particular medium. The mean free path of photons of energy $h\upsilon$ is equal to $1/\mu$, where μ is the linear attenuation coefficient of the medium for photons of energy $h\upsilon$.

The probability is $e^{-\mu x}$ that an x- or γ-ray photon will traverse a slab of thickness x, where $e^{-\mu x}$ is the product of the probabilities that the photon will not interact by any of the interaction processes described earlier. If all modes of interaction except photoelectric absorption, Compton scattering and pair production are negligible, then $e^{-\mu x}$ may be written

$$e^{-\mu x} = (e^{-\tau x})(e^{-\sigma x})(e^{-\kappa x})$$
$$= e^{-(\tau + \sigma + \kappa)x}$$

where τ, σ, and κ represent attenuation coefficients for interaction of the photon by photoelectric absorption, Compton scattering, and pair production, respectively. The total linear attenuation coefficient is:

$$\mu = \tau + \sigma + \kappa$$

Attenuation coefficients vary with the energy of the impinging x- or γ-ray photons and with the atomic number of the absorbing medium. Linear attenuation coefficients also vary with the density of the absorber. Mass attenuation coefficients μ_m, τ_m, σ_m, and κ_m, computed by dividing the linear attenuation coefficients by the density of the medium, do not vary with the density.

$$\mu_m = \frac{\mu}{\rho} \qquad \tau_m = \frac{\tau}{\rho} \qquad \sigma_m = \frac{\sigma}{\rho} \qquad \kappa_m = \frac{\kappa}{\rho}$$

When mass attenuation coefficients are used, the thickness of the absorber should be described in units of grams per square centimeter or milligrams per square centimeter. Total mass attenuation coefficients for air, water, sodium iodide, and lead are plotted in Figure 4-6 as a function of photon energy.

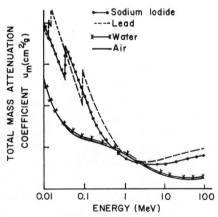

Figure 4-6 Total mass attenuation coefficients μ_m for air, water, sodium iodide, and lead, plotted as a function of photon energy in MeV.

Example 4-5

A narrow monoenergetic beam containing 10,000 photons is reduced to 5000 photons by a slab of aluminum 2 mm thick. What is the total linear attenuation coefficient of aluminum for the photons?

$$I = I_0 e^{-\mu x}$$

$$\frac{I}{I_0} = e^{-\mu x}$$

$$\frac{5000}{10,000} = e^{-\mu(2\,\text{mm})}$$

$$0.5 = e^{-\mu(2\,\text{mm})}$$

$$-0.7 = -\mu(2\,\text{mm}) \quad \text{(See Appendix 1.)}$$

$$\mu = 0.35\,\text{mm}^{-1}$$

The thickness of material required to reduce the intensity* of an x- or γ-ray beam to half is the *half-value layer* (HVL) or *half-value thickness* (HVT) for the beam. The HVL describes the quality or penetrating ability of the beam and is 2 mm Al for the beam described in Example 4-5. The HVL of a monoenergetic narrow beam is

$$\text{HVL} = \frac{\ln 2}{\mu}$$

where μ is the total linear attenuation coefficient of the medium for photons in the beam.

Example 4-6

If 2 mm of aluminum reduces the intensity of the beam described in Example 4-5 to half, what thickness will reduce the intensity to $\frac{1}{8}$ (1250 photons)?

$$I = I_0 e^{-\mu x}$$

$$\frac{I}{I_0} = e^{-\mu x}$$

$$\frac{1250}{10,000} = e^{-(0.35/\text{mm})x}$$

$$0.125 = e^{-(0.35/\text{mm})x}$$

$$-2.1 = -(0.35/\text{mm})x$$

$$x = 6\,\text{mm}$$

If 2 mm aluminum reduces the intensity of the beam to half, then 4 mm reduces the intensity to $\left(\frac{1}{2}\right)\left(\frac{1}{2}\right) = \frac{1}{4}$ and 6 mm reduces the intensity of $\left(\frac{1}{2}\right)\left(\frac{1}{2}\right)\left(\frac{1}{2}\right) = \frac{1}{8}$.

---

*The intensity usually is described as the exposure rate (Chapter 22) or the count rate (Chapter 11).

A narrow monoenergetic beam of γ-ray photons furnishes a curve similar to that in Figure 4-7 when the intensity I is plotted as a function of absorber thickness x. If the data in Figure 4-7 are plotted semilogarithmically (I on a logarithmic scale, x on a linear scale), a straight line is obtained (Figure 4-8). However, the data describe a straight line only if the beam is monoenergetic and if the attenuation measurements are made under conditions of narrow beam geometry. With narrow beam geometry, all scattering processes result in the removal of photons from the beam.

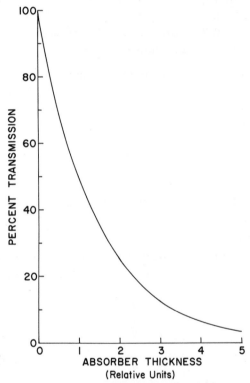

Figure 4-7 Transmission of a narrow monoenergetic beam of photons as a function of absorber thickness.

If the beam is broad, then some photons are scattered from one location in the beam to another. Consequently, the number of photons does not decrease so rapidly in a broad beam as in a narrow beam. The expression for attenuation of a broad beam of photons is

$$I = BI_0e^{-\mu x}$$

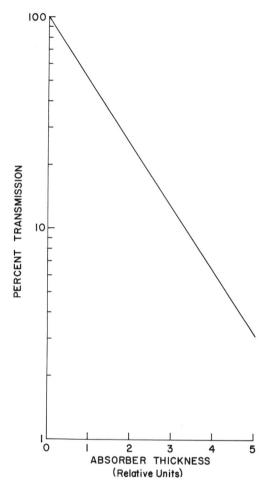

Figure 4-8 Data in Figure 4-7 plotted on a semilogarithmic scale.

where B, the buildup factor, is a number greater than 1. The buildup factor is a complex function of beam size, photon energy, and the geometrical arrangement of source and absorber.

PROBLEMS

1. What is the kinetic energy of an electron ejected from the K orbit of tin ($E_B = 29.1$ keV) by a γ-ray of wavelength 0.2 Å? What is the energy

of the x-ray released as an electron from the L orbit of tin ($E_B = 4.5$ keV) moves to the K orbit?

2. Compute the energy of 662 keV γ-rays from ^{137}Cs which are scattered at 35° and 145° during Compton interactions. What are the energies of the Compton electrons?

3. A γ-ray of 2.75 MeV from ^{24}Na interacts by pair production. What are the kinetic energies of the positron and negatron, if they possess equal energy?

4. A monoenergetic narrow beam of γ-rays is attenuated by copper in the manner shown below. Plot the data semilogarithmically and determine the HVL and total linear attenuation coefficient of the beam.

cpm	6000	1800	545	166	49
thickness, cm	0	1	2	3	4

5. The total mass attenuation coefficient of copper (density = 8.9 g/cm³) is 0.059 cm²/g for 1.0 MeV photons. The number of 1.0 MeV photons in a narrow beam is reduced to what fraction by a slab of copper 2 cm thick?

6. What fraction of a narrow monoenergetic beam remains after it has passed through an absorber of thickness equal to 10 HVL?

7. Compute the minimum and maximum energies of electrons released during Compton scattering of 1.0 MeV γ-rays.

REFERENCES

[1] Hendee, W. R. 1970. Medical radiation physics. Year Book Medical Publishers, Chicago. Chap. 7.

5

Production of Radioactive Nuclides

Although a number of radioactive isotopes are found in nature, most of the isotopes used as radioactive tracers are produced artificially. Methods for producing radioactive isotopes include nuclear fission and the bombardment of selected materials by neutrons or positively charged particles.

Naturally Occurring Radioactive Isotopes

Most radioactive isotopes found in nature are members of one of three radioactive series. Each series consists of sequential transformations beginning with a long-lived parent and ending with a stable nuclide. Intermediate isotopes in the series are in secular equilibrium with the long-lived parent and decay with an apparent half-life equal to that for the parent. All radioactive isotopes found in nature decay by emission of either an α particle or a negatron; consequently, each transformation in a radioactive series changes the mass number of the nucleus either by 4 (α decay) or by 0 (negatron decay).

The uranium series begins with ^{238}U ($T_{1/2} = 4.5 \times 10^9$ yr) and ends with stable ^{206}Pb (Figure 5-1). The mass number of the parent and of each decay product in the series is divisible by 4 with a remainder of 2. Consequently, this series often is referred to as the "$4n + 2$ series", where n represents an integer between 51 and 59. Other radioactive series are the actinium or "$4n + 3$ series" ($^{235}U \rightarrow \rightarrow {}^{207}Pb$)(Figure 5-2) and the thorium or "$4n$ series" ($^{232}Th \rightarrow \rightarrow {}^{208}Pb$)(Figure 5-3). Members of the neptunium or "$4n + 1$ series" ($^{241}Am \rightarrow \rightarrow {}^{209}Bi$) are not found in nature,

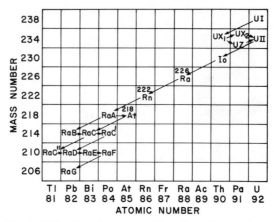

Figure 5-1 Uranium (4n + 2) radioactive series.

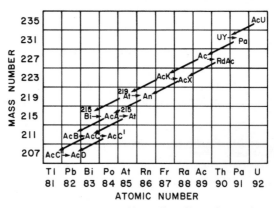

Figure 5-2 Actinum (4n + 3) radioactive series.

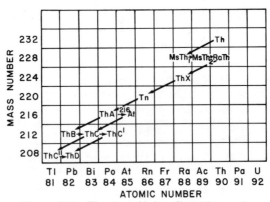

Figure 5-3 Thorium (4n) radioactive series.

because a long-lived parent for this series does not exist. Heavy elements such as those in the radioactive series are seldom present in biologic organisms, and members of the series are used only rarely as radioactive tracers in biologic studies.

A few naturally occurring, long-lived radioactive nuclides are not members of a radioactive decay series. These nuclides include ^{40}K, ^{50}V, ^{87}Rb, ^{115}In, ^{123}Te, ^{138}La, ^{142}Ce, ^{144}Nd, ^{147}Sm, ^{148}Sm, ^{149}Sm, ^{152}Gd, ^{156}Dy, ^{174}Hf, ^{176}Lu, ^{187}Re, ^{190}Pt, and ^{204}Pb.

Nuclear Reactors

In a nuclear reactor, a fissionable nuclear fuel such as ^{235}U or ^{239}Pu is placed in a defined geometry within the reactor core. Fast-moving neutrons released as the ^{235}U or ^{239}Pu nuclei fission are slowed to thermal energy (~ 0.025 eV) by a low-Z moderator such as water, heavy water, beryllium, or graphite surrounding or mixed intimately with the nuclear fuel. The slowly moving neutrons are absorbed by additional ^{235}U or ^{239}Pu nuclei, and a chain reaction is maintained. The rate of absorption of slow neutrons by fissionable nuclei determines the fission rate for the reactor, and is controlled by placing inert neutron absorbers in the reactor core. Typical neutron absorbers, usually termed *control rods*, include boron and cadmium. Heat generated in the reactor core is removed by a coolant such as water or one of a variety of gases or liquid metals.

Many uses have been found for nuclear reactors, including the production of nuclear power, the production of nuclear fuel, the testing of materials exposed to large amounts of radiation, and the production of radioactive isotopes. Most of the major countries have reactors suitable for producing radioactive isotopes.

Isotope Production by Neutron Bombardment

Most radioactive isotopes used in biologic studies are produced in a nuclear reactor by neutron bombardment of a selected target. Typical reactions may be categorized as (n,γ), (n,p), (n,α), or [(n,γ) $\rightarrow$ daughter] reactions, or as multistage processes.

B (n,γ)C Reaction.   A target nucleus B bombarded with neutrons (often thermal neutrons) may be transformed into a product nucleus C with the emission of a γ-ray photon.

$$_Z^A B + {}_0^1 n \rightarrow {}_Z^{A+1} C + \gamma$$

The common notation for this type of reaction is

$$B(n,\gamma)C$$

and typical reactions are

$$^{23}_{11}Na(n,\gamma)^{24}_{11}Na \qquad ^{197}_{79}Au(n,\gamma)^{198}_{79}Au$$

$$^{31}_{15}P(n,\gamma)^{32}_{15}P \qquad ^{46}_{20}Ca(n,\gamma)^{47}_{20}Ca$$

In the (n,γ) reaction, the target and product nuclei are isotopes of the same element, and usually cannot be separated chemically. Consequently, radioactive samples prepared by an (n,γ) reaction usually exhibit limited specific activity. Occasionally, the specific activity of these samples may be increased by a process such as a Szilard–Chalmers reaction.*

B(n,p)C Reaction.   A target nucleus B bombarded with fast neutrons may be changed into a nucleus C with the emission of a proton.

$$^{A}_{Z}B + ^{1}_{0}n \rightarrow {}_{Z-1}^{A}C + ^{1}_{1}p$$

The common notation for this type of reaction is

$$B(n,p)C$$

and typical reactions are

$$^{35}_{17}Cl(n,p)^{35}_{16}S \qquad ^{32}_{16}S(n,p)^{32}_{15}P$$

$$^{14}_{7}N(n,p)^{14}_{6}C \qquad ^{58}_{28}Ni(n,p)^{58}_{27}Co$$

The product of an (n,p) reaction differs chemically from the target element, and the two may be separated chemically. Consequently, radioactive samples of high specific activity may be obtained with the (n,p) reaction.

B(n,α)C Reaction.   A target nucleus B bombarded with fast neutrons may receive enough energy to eject an α particle and form a product C which differs chemically from the target nuclide.

$$^{A}_{Z}B + ^{1}_{0}n \rightarrow {}_{Z-2}^{A-3}C + ^{4}_{2}\alpha$$

Typical (n,α) reactions include

$$^{27}_{13}Al(n,\alpha)^{24}_{11}Na$$

$$^{6}_{3}Li(n,\alpha)^{3}_{1}H$$

B(n,γ)C $\rightarrow$ D Reaction.   If the product C of an (n,γ) reaction decays to a radioactive daughter D, then the daughter can be separated chemically

---

*In a Szilard–Chalmers reaction, products of an (n,γ) reaction are in a different chemical form than the target atoms because chemical bonds are broken as the atom recoils during transformation.

from other elements in the target, and a sample of high specific activity can be obtained.

$$\tfrac{A}{Z}B(n,\gamma)^{A+1}_{\ \ Z}C \xrightarrow{\text{decay}} D$$

Typical reactions include

$$^{130}_{52}Te(n,\gamma)^{131}_{52}Te \xrightarrow[\ ]{T_{1/2}=25\ min} {}^{131}_{53}I + {}^{\ 0}_{-1}\beta + v$$

$$^{124}_{54}Xe(n,\gamma)^{125}_{54}Xe \xrightarrow[\text{electron capture}]{T_{1/2}=17\ hr} {}^{125}_{53}I + v$$

Multistage Processes. Sometimes a radioactive nuclide is produced by successive capture of two or more neutrons, possibly with one or more of the products undergoing β decay. For example, a target nuclide B might be transformed into the nuclide E by the process

$$B(n,\gamma)C(n,\gamma)D \xrightarrow{\text{decay}} E$$

For example, ^{241}Am is produced from ^{239}Pu by

$$^{239}_{94}Pu(n,\gamma)^{240}_{94}Pu(n,\gamma)^{241}_{94}Pu \xrightarrow{T_{1/2}=13.2\ yr} {}^{241}_{95}Am + {}^{\ 0}_{-1}\beta + v$$

Multistage processes are utilized often to produce transuranic elements (elements with atomic numbers greater than 92).

Equations for Isotope Production

The production of a radioactive isotope by an (n,γ), (n,p), or (n,α) reaction may be described mathematically by a rather simple expression. If the number of target nuclei remains essentially constant, if the product nuclide is a poor absorber of neutrons compared to the target, and if no product nuclei are present at time $t = 0$, then the activity of a sample bombarded for time t is

$$A = \frac{\phi N\sigma}{3.7 \times 10^{10}}(1 - e^{-\lambda t})$$

where ϕ = neutron flux in neutrons per square centimeter per second

N = number of target nuclei in the sample

σ = absorption cross-section of target nuclei in square centimeters (the absorption cross-section or "cross-sectional area presented to a neutron by a target nucleus" usually is described in units of barns, with 1 barn = 10^{-24} cm^2)

λ = decay constant of the product isotope ($\lambda = 0.693/T_{1/2}$, where $T_{1/2}$ is the half-life of the product)

A = activity of the sample in curies

When the bombardment time is greater than many half-lives of the product, the equation above reduces to

$$A_{\text{Satn}} = \frac{\phi N \sigma}{3.7 \times 10^{10}}$$

where A_{Satn} represents the maximum or "saturation" activity of the product in curies. In Figure 5-4, the buildup of ^{24}Na activity is demonstrated for natural sodium irradiated by thermal neutron fluxes of 2×10^{12} and 3×10^{12} neutrons/cm²-sec. As shown in Figure 5-4, the activity of the sample is half of its saturation activity after an irradiation time of one half-life of the product, three-fourths of its saturation activity after an irradiation time of two half-lives, etc. Samples seldom are irradiated for more than two or three half-lives of the product, because the rate of increase in activity diminishes rapidly with irradiation time. To increase the specific activity of a product, a sample should be subjected to a greater flux of neutrons rather than to a longer irradiation time.

The equations above for sample activity are not applicable to situations where the decrease in target nuclei is not negligible or where the product nuclide absorbs a significant number of neutrons. In these situations, saturation activity is not achieved, and the sample activity increases initially to a maximum value and decreases thereafter. The maximum activity may be considerably less than the saturation activity predicted by the equation above. For example, a saturation specific activity of about 1000 Ci/g is predicted by the equation for a sample of ^{59}Co exposed to a

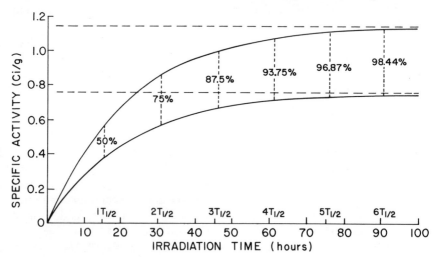

Figure 5-4 Buildup of ^{24}Na activity in a sample of natural sodium bombarded with thermal neutron fluxes of 2×10^{12} n/(cm²-sec) (lower curve) and 3×10^{12} n/(cm²-sec) (upper curve).

neutron flux of 10^{14} n/cm²-sec (Example 5-1). If the reduction in target nuclei and the activation of product nuclei are considered, then a maximum activity of only 370 Ci/g is achieved after about 7 years of irradiation[1].

Example 5-1

Estimate the specific activity after 1 year and the saturation specific activity for a 1 g sample of ^{59}Co exposed to 10^{14} n/cm²-sec. The activation cross-section is 37 barns for the reaction ^{59}Co(n,γ)^{60}Co, and the half-life of ^{60}Co is 5.3 years.

The number of atoms N in 1 g of elemental cobalt is

$$N = \frac{(1 \text{ g})(6.02 \times 10^{23} \text{ atoms/g-atomic mass})}{59 \text{ g/g-atomic mass}}$$

$$= 1.0 \times 10^{22} \text{ atoms}$$

The predicted activity of the 1 g mass after 1 year of irradiation is

$$A = \frac{\phi N \sigma (1 - e^{-\lambda t})}{3.7 \times 10^{10}}$$

$$= \frac{(10^{14} \text{ n/cm}^2\text{-sec})(1.0 \times 10^{22} \text{ atoms})(37 \times 10^{-24} \text{ cm}^2)}{3.7 \times 10^{10} \text{ atoms/sec-Ci}}$$

$$\times (1 - e^{-0.693(1)/5.3})$$

$$= 124 \text{ Ci/g}$$

The predicted saturation specific activity A_{Satn} is

$$A_{\text{Satn}} = \frac{\phi N \sigma}{3.7 \times 10^{10}}$$

$$= \frac{(10^{14} \text{ n/cm}^2\text{-sec})(1.0 \times 10^{22} \text{ atoms})(37 \times 10^{-24} \text{ cm}^2)}{3.7 \times 10^{10} \text{ atoms/sec-Ci}}$$

$$= 1000 \text{ Ci/g}$$

If target "burnup" and product activation are considered, then the specific activity after 1 year is 106 Ci/g and the maximum activity attainable is 370 Ci/g.

Fission Products

Residual radioactive nuclei are produced when a nucleus of ^{235}U or ^{239}Pu fissions, and these nuclei and their radioactive daughters furnish a wide variety of radioactive fission products which may be separated during the recovery of unused ^{235}U or ^{239}Pu from "spent" nuclear fuel. The percent fission yield (the number of specific byproduct nuclei per 100 nuclear fissions) is plotted in Figure 5-5 for ^{233}U, ^{235}U, and ^{239}Pu. The

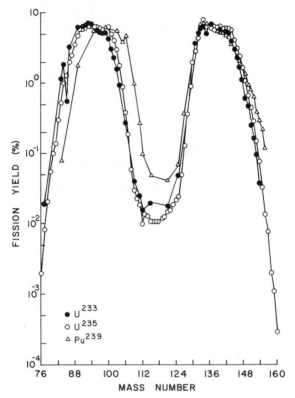

Figure 5-5 Percent fission yield as a function of mass number
for products obtained by fission of ^{233}U, ^{235}U, and ^{239}Pu.

fission products depicted in Figure 5-5 include many isotopes of biologic
interest such as ^{85}Kr, ^{90}Sr, ^{99}Mo, ^{106}Ru, ^{131}I, ^{132}I, ^{133}Xe, and ^{137}Cs. The
notation for this method of producing radioactive isotopes is

$$U(n,f) \rightarrow C,$$

where C is the desired radioactive product.

Isotope Production by Charged-Particle Bombardment

A variety of devices have been developed for the acceleration of charged
particles to high energy. The accelerator used most widely to produce
radioactive isotopes by charged-particle bombardment is the cyclotron.
The operation of a cyclotron is depicted in Figure 5-6. Two hollow, semi-
circular electrodes or "dees" are mounted between the poles of an elec-

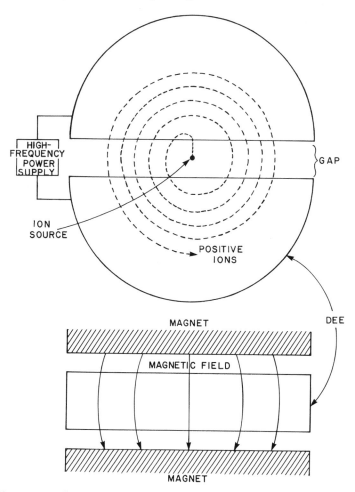

Figure 5-6 Conventional cyclotron. The path of the accelerated particles is denoted by the dashed curve. The particles are accelerated each time they cross the gap between the dees.

tromagnet and separated from each other by a gap of 1 or 2 in. The electromagnet is energized by direct current and furnishes a magnetic field of constant intensity across the dees. An alternating voltage with a frequency of 10–40 megahertz (MHz)* is applied across the dees.

Positive ions (for example, $^1H^+$, $^2H^+$, $^3He^{2+}$ or $^4He^{2+}$) are released by a cathode-arc source in the gap between the dees and are accelerated in

---

*1 megahertz (MHz) = 10^6 hertz (Hz), where the hertz is a unit of frequency equal to 1 cycle per second.

bunches toward the negative dee. As the particles enter the hollow dee, they are shielded from the electric field. The magnetic field forces the particles into a circular path which they follow with constant speed. As the polarity reverses across the dees, the ions emerge from the dee and accelerate across the gap into the opposite dee. The ions follow a circular orbit in this dee also, and emerge again just as the polarity reverses. In this manner, the particles are accelerated each time they cross the gap between the dees until they obtain the desired energy. For example, deuterons ($^2H^+$) may be accelerated to an energy of 35 MeV in a conventional cyclotron.

After the accelerated particles have attained the desired energy, they are directed onto a target chosen for the production of a particular isotope. Typical reactions include those listed below:

$$^{10}B(d,n)^{11}C \qquad ^{24}Mg(d,\alpha)^{22}Na$$
$$^{14}N(d,n)^{15}O \qquad ^{12}C(p,3p3n)^7Be$$
$$^{50}Cr(\alpha,2n)^{52}Fe \qquad ^{31}P(\alpha,n)^{34}Cl$$
$$^{52}Cr(\alpha,4n)^{52}Fe \qquad ^{48}Ca(p,pn)^{47}Ca$$
$$^{51}V(p,n)^{51}Cr \qquad ^{16}O(\alpha,pn)^{18}F$$

Isotopes produced by neutron bombardment or nuclear fission almost always have a ratio of neutrons to protons greater than that required for stability. Consequently, these isotopes generally decay by emission of a negatron. Isotopes produced by charged-particle bombardment usually are deficient in neutrons and decay by electron capture and, frequently, positron emission.

Isotope production in a cyclotron almost always involves the transmutation of one element into another. Consequently, the radioactive product of a charged-particle reaction usually can be separated easily from the target.

Selection of a Target Material

The target chosen for neutron or charged-particle bombardment depends upon a number of considerations, including possible damage to the reactor or accelerator, the safety of persons handling the activated target, the conditions required for subsequent chemical processing, the purity and specific activity required for the product, and the cost of the target material. Usually, the target must be cooled during irradiation, and must be inserted and recovered while the reactor or accelerator is operating. The target should be as stable as possible against the combined effects of heating and irradiation, and should be enclosed within a sealed container if corrosive products, gases, or vapors are released during irradiation. In general, target materials in elemental form or in the form of oxides or simple anhydrous inorganic salts are desirable for bombardment by neutrons or charged particles. The amount of target material to be irradiated often is limited by the level of radioactivity which can be handled safely and, sometimes in reactors, the effect that absorption of neutrons by the target has upon the availability of neutrons for maintenance of reactor criticality.

Almost always, an irradiated target must be processed chemically to obtain the product isotope in the chemical form desired and with the highest purity. The requirements for chemical processing should be considered when the target is chosen. The purity of the target also must be considered, because activated impurities may contribute unwanted radioactivity and cause difficulties in handling the irradiated target. Also, impurities in the target which are isotopes of the desired product may reduce the product's specific activity. For example, a trace of natural cobalt (^{59}Co) in natural nickel furnishes an unacceptable level of ^{60}Co activity in ^{58}Co produced by neutron bombardment of the nickel. Since the ^{60}Co and ^{58}Co cannot be separated chemically, the only practical method for reducing the ^{60}Co activity is selection of a nickel target which contains as little cobalt as possible. Similarly, when nickel is used to prepare ^{57}Co according to the reaction

$$^{58}\text{Ni(p,pn)}^{57}\text{Ni} \xrightarrow{\; T_{1/2} = 36.1 \text{ hr} \;} {}^{57}\text{Co}$$

where the decay of ^{57}Ni to ^{57}Co is accomplished by electron capture and positron emission, nonradioactive ^{59}Co may be produced by the reactions

$$^{60}\text{Ni(p,2p)}^{59}\text{Co}$$
$$^{62}\text{Ni(p,}\alpha\text{)}^{59}\text{Co}$$

The ^{59}Co cannot be separated chemically from ^{57}Co, and the specific activity of the ^{57}Co product is reduced by the stable ^{59}Co. This problem may be avoided only by using pure ^{58}Ni as the target material.

As another example, neutron bombardment of natural mercury produces ^{197}Hg ($T_{1/2} = 65$ hr) contaminated with ^{203}Hg ($T_{1/2} = 47$ days). As the short-lived ^{197}Hg decays after irradiation, the percent contamination of the sample by ^{203}Hg increases. The reactions are

$$^{196}\text{Hg(n,}\gamma\text{)}^{197}\text{Hg}$$
$$^{202}\text{Hg(n,}\gamma\text{)}^{203}\text{Hg}$$

The presence of the long-lived ^{203}Hg may be objectionable, particularly in studies where ^{197}Hg is delivered to humans. To reduce the ^{203}Hg contaminant, a target enriched in ^{196}Hg must be used. For example, electromagnetic separation techniques may be employed to increase the amount of ^{196}Hg in the target from its natural abundance of 0.146% to about 25%. Irradiation of a target enriched to 25% with ^{196}Hg provides ^{197}Hg essentially uncontaminated with ^{203}Hg.

During an (n,γ) reaction, the nucleus and atom recoil with a force which may disrupt the bonding of the atom within the target molecular structure. After recoil, the atom may return to its original state of bonding, or it may recombine differently to produce an altered molecular configuration. This effect, termed the *Szilard–Chalmers effect*, is utilized occasionally to obtain a radioactive isotope with enhanced specific activity. Whether or not the Szilard–Chalmers effect is utilized for separation of a radioactive product, its influence on the chemical state of the product should be considered when a target material is chosen.

Chemical Processing of an Irradiated Target

Undesired radioactive products may be produced by activation of impurities in an irradiated target. For example, suppose that a target contains an element B and an impurity element C with a higher cross-section for the (n,γ) reaction. The impurity C cannot be reduced to less than 0.1% by weight of B. Suppose that because of its higher cross-section the activity obtained by irradiating C contributes 10% of the total activity of the sample. After irradiation, 100 mg of nonradioactive C are added for every gram of B in the sample. This procedure is followed by a chemical separation to return the concentration of C to 0.1% by weight of B. The contribution of C to the total activity of the sample will be reduced from 10 to 0.1% by the chemical separation. If the separation is repeated, the contribution of C can be reduced further from 0.1 to 0.001%.

Isotopes are produced in small quantities (usually a few micrograms or less, see Table 5-1) by neutron or charged-particle bombardment, and often an inactive carrier must be added before employment of conventional processes for separation (for example, fractional distillation, fractional crystallization, or precipitation). However, if the greatest

TABLE 5-1 *Mass and Activity of Products Produced by Exposure of a Target for 1 Week to a Neutron Flux Density of $10^{12}n/(cm^2\text{-}sec)$*

Target element	Reaction	Product	Mass, g	Activity, mCi
Barium	(n,γ) $\xrightarrow{\text{decay}}$	^{131}Cs	2×10^{-9}	0.2
Nickel	(n,p)	^{58}Co	3×10^{-9}	0.1
Chlorine	(n,p)	^{35}S	1.5×10^{-7}	6.3

specific activity possible is desired for an isotope, then a carrier should not be added. Instead, a method for separating very small quantities of material should be employed. A few techniques which may be applicable to particular situations are outlined here[1].

Filtration. Some carrier-free isotopes can be removed from neutral solution by passage through a paper or sintered glass filter. For example, the decay product ^{47}Sc can be removed from ^{47}Ca by filtration of a solution containing the two nuclides. The selective filtration of ^{47}Sc probably is related to absorption of hydrated scandium ions on the filter surface.

Absorption and Coprecipitation. Many impurities, both cations and anions, can be scavenged from solution by addition of Fe^{3+} followed by precipitation of $Fe(OH)_3$. For example, carrier-free isotopes of manganese and cobalt may be removed from solution on a ferric hydroxide precipitate. The iron can be removed from the final product by ether extraction from solution in 6 N HCl, or by passage through an anion exchange column.

Solvent Extraction. Many high specific activity isotopes may be prepared by solvent extraction, sometimes with the addition of specific organic complexing agents. For example, ^{45}Ca may be extracted from neutron-irradiated scandium in HCl solution by adding thenoyl trifluoracetone in benzene. Similarly, ^{90}Sr may be extracted from a solution containing fission products by adding a solution of sodium di(2-ethyloxyl)-phosphate.

Chromatography. Chromatographic techniques, particularly ion exchange chromatography, are important in the separation of carrier-free isotopes from irradiated targets. For example, some rare earths may be separated on cation exchange columns if an appropriate complexing eluant is used, and metals in hydrochloric acid solution often can be separated on an anion exchange resin[2].

A few special techniques for separating isotopes from irradiated targets are discussed briefly in *The Radiochemical Manual*[1].

Radionuclide Generators

Radionuclide generators furnish short-lived radioactive isotopes which otherwise might be unavailable because of the decay of the isotopes during transport from a remote site. In a generator, a radioactive isotope with a relatively long half-life is attached firmly to an ion-exchange column. The isotope decays to a radioactive daughter with a shorter half-life. The daughter is an isotope of a different element and is attached only loosely to the column. Consequently, the daughter may be removed with an eluant such as dilute ammonium hydroxide, hydrochloric acid, or saline water. If the daughter is removed by elution ("milking"), it re-accumulates on the column and rapidly attains a state of transient equilibrium with the longer lived parent. In Figure 5-7, activities of

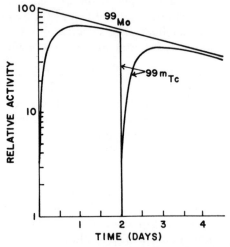

Figure 5-7 Activity of parent and daughter in a ^{99}Mo–^{99m}Tc generator. The ^{99m}Tc activity remains below the activity of ^{99}Mo because 14% of the ^{99}Mo nuclei decay directly to ^{99}Tc rather than to ^{99m}Tc. The ^{99m}Tc is removed from the generator at time = 2 days.

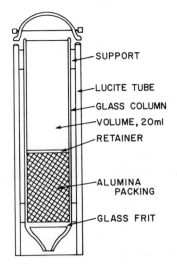

Figure 5-8 Typical configuration of a radionuclide generator.

parent ^{99}Mo and daughter ^{99m}Tc are depicted for a ^{99}Mo–^{99m}Tc generator or "cow." A few useful and potentially useful nuclides available from radionuclide generators are listed in Table 5-2.

A radionuclide generator should furnish a product in a chemical form which requires little preparation before use. Furthermore, the product should not be contaminated by undesired radioactive or stable isotopes. To reduce difficulties in shipping and shielding, the generator should be rugged and compact. The ion-exchange resin should not decompose during irradiation. A typical generator configuration is diagrammed in Figure 5-8.

TABLE 5-2 *Gamma-Emitting Radioactive Nuclides Available by Eluting an Ion-Exchange Column Containing a Longer Lived Parent*

Daughter	Daughter half-life	Decay	Photon energy, keV	Parent	Parent half-life
^{28}Al	2.3 min	β^-	1780	^{28}Mg	21.3 hr
^{38}Cl	37 min	β^-	2170	^{38}S	2.9 hr
^{42}K	12.4 hr	β^-	1520	^{42}Ar	33 yr
^{44}Sc	3.9 hr	β^+, EC	511, 1160	^{44}Ti	47 yr
^{68}Ga	68 min	β^+, EC	511	^{68}Ge	287 days
^{72}As	26 hr	β^+, EC	511, 840, 638, others	^{72}Se	8.5 days
^{82m}Rb	6.3 hr	β^+, EC	780	^{82}Sr	25 days
^{83m}Kr	1.9 hr	IT	32	^{83}Rb	83 days
^{87m}Sr	2.8 hr	IT	388	^{87}Y	80 hr
^{99m}Tc	6 hr	IT	140	^{99}Mo	67 hr
^{103m}Rh	57 min	IT	40	^{103}Ru	40 days
^{111m}Cd	49 min	IT	150	^{111}Ag	7.5 days
^{113m}In	1.7 hr	IT	393	^{113}Sn	115 days
^{115m}In	4.5 hr	IT	340	^{115}Cd	2.2 days
^{123}I	13.1 hr	EC	159	^{123}Xe	1.85 hr
^{125m}Te	58 days	IT	110, 35	^{125}Sb	2.7 hr
^{127}Te	9.3 hr	β^-	418, others	^{127m}Te	109 days
^{131}Cs	9.7 days	EC	29	^{131}Ba	11.7 days
^{132}I	2.3 hr	β^-	960, 780 670, 530	^{132}Te	3.2 days
^{137m}Ba	2.6 min	IT	662	^{137}Cs	30 yr
^{144}Pr	17.3 min	β^-	2180, 1490, 694	^{144}Ce	284 days
^{189m}Os	5.9 hr	IT	31	^{189}Ir	13 days
^{194}Au	1.6 days	β^+, EC	328	^{194}Hg	1.3 yr

Radionuclide generators are used extensively for preparation of nuclides such as ^{99m}Tc, ^{113m}In, ^{87m}Sr, and ^{68}Ga used in diagnostic medicine. For laboratory experiments involving short-lived isotopes, "mini-generators" are available commercially for the production of ^{137m}Ba and ^{113m}In.

PROBLEMS

1. A 1 g sample of natural silver (51.3% ^{107}Ag, 48.7% ^{109}Ag) is irradiated with a thermal neutron flux density of 8×10^{12} n/cm^2-sec. The cross-section for the reaction ^{109}Ag (n,γ) ^{110m}Ag is 1.56 barns, and the half-life of ^{110m}Ag is 253 days. Assuming negligible target burnup and product activation, compute the ^{110m}Ag activity after 4 weeks of irradiation.

2. Complete the reactions below:

   (a) $^{70}_{32}$Ge(α,2n)____

   (b) $^{102}_{44}$Ru(n,γ)____

   (c) $^{81}_{35}$Br(α,___)$^{84}_{37}$Rb

   (d) $^{181}_{73}$Ta(___,2n)$^{183}_{75}$Re

   (e) $^{40}_{18}$Ar(___,p)$^{43}_{19}$K

   (f) $^{33}_{16}$S(n,p)____

   (g) $^{209}_{83}$Bi(___,2n)$^{208}_{84}$Po

   (h) ____(d,n)$^{15}_{8}$O

   (i) $^{31}_{15}$P(___,γ)$^{32}_{15}$P

   (j) $^{12}_{6}$C(___,n)$^{13}_{7}$N

   (k) $^{238}_{92}$U(n,___)$^{237}_{92}$U $\xrightarrow{\beta^-}$____

   (l) $^{37}_{17}$Cl(___,6p4n)$^{28}_{12}$Mg

3. Write the following reactions for the bombardment of $^{32}_{16}$S with neutrons: (n,p), (n,α), (n,d), (n,γ)

4. A 20 g sample of stable phosphorus (^{31}P) is bombarded with neutrons until the activity of the ^{32}P in the sample is 2 mCi. What is the ratio of radioactive to stable atoms in the sample?

5. A 1 cm^2 sheet of gold foil 2×10^{-3} cm thick (density $= 19$ g/cm^3) is exposed for 10 min to a thermal neutron flux density of 5×10^{10} n/cm^2-sec. After irradiation, the activity of ^{198}Au ($T_{1/2} = 2.7$ days) produced by an (n,γ) reaction is 5×10^9 disintegrations/hr. What is the activation cross-section of ^{197}Au for thermal neutrons?

6. Write three reactions in which targets composed of stable nuclides may be used to produce (a) ^{14}C, (b) ^{15}O, (c) ^{60}Co.

REFERENCES

[1] Wilson, B. (ed.). 1966. The radiochemical manual, 2nd ed. The Radiochemical Centre, Amersham, England.

[2] Kraus, K., and F. Nelson. 1956. Symposium on ion-exchange and chromatography in analytical chemistry, Atlantic City, N.J., June, 1956. ASTM Spec. Tech. Pub. 195, Philadelphia. p. 27.

6

Synthesis of Labeled Compounds

A labeled compound contains an uncommon atom or group of atoms which permits differentiation between the labeled compound and the compound in its more common form. The uncommon component or "label" may be a stable isotope (for example, ^{15}N in place of ^{14}N) or a chemical group (for example, CH_3 in place of H). More often, however, the label is a radioactive isotope, usually an isotope of an element which occurs naturally in biologic organisms. That is, the most common labels are radioactive isotopes of hydrogen, carbon, sulfur, phosphorus, and iodine. Radioactive isotopes of nitrogen and oxygen are not available with half-lives long enough for common use of these isotopes as labels.

Isotopic and Nonisotopic Labeling

A compound may be labeled isotopically or nonisotopically. An isotopically labeled compound is a compound containing one or more radioactive atoms which have replaced stable atoms of the same element in the compound. The replacement occurs with no other change in the compound; consequently, an isotopically labeled compound mimics the chemical and biological properties of its nonlabeled counterpart. An example of an isotopically labeled compound is propionic acid in which ^{14}C has replaced one or more stable carbon atoms (for example, $^{14}CH_3CH_2COOH$).

A nonisotopically labeled compound is a compound which has been labeled by a procedure more complex than simple substitution of a radioactive atom for its stable counterpart. For example, the labeling of human serum albumin with ^{125}I or ^{131}I is nonisotopic, because iodine is not a natural constituent of albumin. In general, results of studies with nonisotopically labeled compounds must be interpreted with caution, because the addition of the label may change the chemical and biological properties of the compound.

Specific and General Labeling

A compound which is labeled specifically contains a radioactive atom at only a single specified position in the molecule. Almost always, a specifically labeled sample is a mixture of unlabeled molecules and molecules with a radioactive atom in the single specified position. For example, a sample of propionic acid is labeled specifically with ^{14}C if all the labeled molecules have the ^{14}C atom in the same position, even though many unlabeled molecules may be present in the sample.

A radioactive compound which is not labeled specifically is said to be randomly labeled or generally labeled. In these compounds, the position of the radioactive atom in the molecule varies from one molecule to the next. If the compound is labeled uniformly, then the radioactive atom appears with statistical uniformity at each available site within the molecule. If the radioactive atom appears with greater frequency at some sites and with lesser frequency at others, then the compound is said to be labeled nonuniformly.

Exchange, Substitution, and Addition Reactions

In general, exchange, substitution, and addition reactions are among the easier procedures for adding a radioactive atom to a compound. For example, one of these reactions is almost always employed with tritium gas or tritiated water to form a tritium-labeled intermediate to be used for chemical synthesis of a desired product[1]. The tritium-labeled intermediate may be produced by:

(1) Exchange reaction catalyzed by a heterogeneous catalyst
(2) Exchange reaction catalyzed by a homogeneous catalyst
(3) Exchange reaction catalyzed by radiation
(4) Substitution reaction achieved by chemical reduction
(5) Addition (hydrogenation) reaction

Exchange with Heterogeneous Catalysis. This method is perhaps the most rapid and least difficult procedure for labeling a compound with tritium. An unlabeled organic compound is heated in the presence of tritium and a catalyst such as palladium or platinum. After labeling has been achieved, labile tritium is removed by repeated equilibration with water or other solvent and the product is purified. Labeling of the compound is random but rarely uniform, and high specific activities (1–20 Ci/mM)* are attainable. Exchange reactions catalyzed by a heterogeneous catalyst are used frequently for compounds which are stable under the conditions required for labeling.

Exchange with Homogeneous Catalysis. Hydrogen atoms in labile positions in a molecule may be replaced easily by tritium atoms available in an environment of tritiated water (^3H_2O). For example, malonic acid may be labeled with tritium simply by mixing the acid with tritiated water:

$$CH_2(COOH)_2 + 3\,^3H_2O \rightleftharpoons CH^3H(COO^3H) + 3H\,^3HO$$

When dissolved in water, compounds labeled by this procedure rapidly lose tritium atoms by exchange with hydrogen in the water. Consequently, the products of a simple exchange reaction usually must be processed further before a useful product is obtained. For example, malonic acid labeled with tritium may be decarboxylated to acetic acid containing ^3H in the methyl group:

$$CH^3H(COO^3H)_2 \xrightarrow{\text{heat}} CH^3H_2COOH + CO_2$$
$$\downarrow{\text{+NaOH}}$$
$$\hookrightarrow CH^3H_2COONa + H^3HO$$
$$\hookrightarrow CH^3H_2COOH$$

Exchange Catalyzed by Radiation (the Wilzbach Method[2]). Many compounds may be labeled by exposure for a few days or more to an atmosphere of a few curies of tritium gas[3]. The exchange reaction is catalyzed by energy released during the decay of tritium atoms. Additional energy may be supplied by an electric discharge in the presence of the sample or by exposure of the sample to ultraviolet, γ, or microwave radiation[4]. Some exchange of tritium for hydrogen occurs with almost all compounds subjected to this procedure; however, the usefulness of the procedure is reduced by side effects such as the addition of tritium to unsaturated centers and the radiation-induced degradation of the compounds. The latter effect limits the specific activity attainable by exchange reactions catalyzed by radiation, and also necessitates rigorous purification and analysis of the product. For these reasons, compounds usually are labeled by radiation catalysis only if no alternate method is available.

*The specific activity of a labeled compound is the activity per unit mass, expressed in units of Ci/mM, mCi/mg, etc.

Substitution by Chemical Reduction. In the presence of tritiated water or tritium-labeled acetic acid, a halogenated compound RX may be reduced by a metal M (for example, zinc) to furnish a product with tritium in place of the halogen atom:

$$RX + M^{2+} + {}^3H_2O \rightarrow R^3H + MO^3HX$$

The large amount of tritium required for this process may be reduced by converting the halogenated compound into an organometallic compound such as a Grignard reagent (RMgX), which subsequently is reacted with tritiated water:

$$RMgX + {}^3H_2O \rightarrow R^3H + MgO^3HX$$

Halogenated compounds also have been reduced with tritium gas and a catalyst such as palladium or platinum:

$$RX + {}^3H_2 \xrightarrow{\text{Pd or Pt}} R^3H + {}^3HX$$

Similar methods, using sodium borotritide or tritiated lithium borohydride with omission of the metal catalyst, are used often to label carbonyl compounds and carbohydrates[1, 3].

Except for the procedures which involve a heterogeneous catalyst, substitution reactions by chemical reduction usually furnish specifically labeled products.

Addition (Hydrogenation). A useful method for tritium labeling involves the addition of tritium to unsaturated centers (particularly carbon–carbon double and triple bonds) in a molecule[5, 6]. The method requires a non-hydroxylic solvent and a palladium or platinum catalyst, and usually furnishes a product with tritium at locations other than unsaturated centers. High specific activities are attainable with this method of labeling.

Chemical Synthesis

One or more radioactive atoms may be added to a compound during chemical synthesis of the compound[4, 7, 8]. By adding the label as the compound is synthesized, some control may be exercised over yield, position of the label, and purity of the product. Usually, chemical synthesis is the preferred method for labeling compounds with ^{14}C, ^{32}P, ^{35}S, ^{131}I, and ^{125}I. For ^{14}C labeling, the synthesis often begins with a one-carbon compound such as $^{14}CO_2$ and proceeds to the desired product through an intermediate such as benzene or succinic acid.

Chemical synthesis usually furnishes a specifically labeled compound of relatively low specific activity. This method often provides a racemic mixture of D- and L-isomers of the labeled compound, and some difficulties may be encountered when the mixture is used in biologic studies.

Biosynthesis

Various living organisms and enzyme systems have been proposed for biosynthesis of labeled compounds. However, not all of the proposed organisms and enzyme systems synthesize and accumulate sufficient amounts of the desired compounds with adequate specific activity. Drawbacks to biosynthetic methods include difficulties in separating and purifying the labeled compounds and in determining the patterns of labeling. Compounds labeled by biosynthesis include many macromolecules (for example, nucleic acids and proteins) as well as simpler molecules (for example, vitamins, hormones, and sugars).

Many compounds may be labeled with ^{14}C by photosynthesis. For example, labeled starch, glucose, fructose, and sucrose are provided by detached plant leaves exposed to $^{14}CO_2$ and light. High specific activities and uniform labeling often are attainable by photosynthesis. Proteins, amino acids, and lipids labeled with ^{14}C or ^3H have been isolated from the green algae *chlorella* exposed to a $^{14}CO_2$ atmosphere or to tritiated water.

"Isotope farming" involves the isolation and purification of labeled compounds from intact plants grown to maturity in a $^{14}CO_2$ atmosphere. Although this method appears promising for large-scale production of labeled compounds, a number of problems have been encountered. For example, the specific activity and yield are very low for most compounds, primarily because the radioactivity must be maintained at a relatively low concentration, (for example 1 mCi/g of $^{14}CO_2$ or 20 mCi/ml of tritiated water) to prevent the radiation from inhibiting the growth of the plants[9]. The production of isomers which are preferred biologically is one advantage of biosynthetic labeling.

Hot Atom Chemistry

The preparation of ^{14}C-labeled compounds by "hot atom chemistry" involves the bombardment of a sample with neutrons to produce a ^3H- or ^{14}C-labeled product[10]. This method of labeling is of limited practical significance, however, because the yield and specific activity are low for most products, and also because rigorous purification of the products is required.

Purity of Labeled Compounds

The purity of a radioactive sample may be described in a number of ways. The radiochemical purity of a sample is the proportion of the total sample activity which is present in the desired chemical form. The desired chemical form may include specification of the optical configuration desired for the compound and the position of the radioactive label within the compound. The radiochemical purity may be determined by various methods, including paper or thin-layer chromatography, isotope dilution analysis, and ultraviolet or infrared absorption spectroscopy.

The radiochemical purity required for a particular compound varies with its intended use. For example, a high degree of radiochemical purity is not required for ^{14}C-labeled sodium acetate used in the biosynthesis of more complex compounds because the organism performing the biosynthesis will select the desired compound from an impure mixture. Greater radiochemical purity is required when preference for the desired compound is less pronounced, or when the results of an experiment may be influenced by the presence of small amounts of an impurity. The position of the radioactive label in a compound often is unimportant when the compound remains intact during the course of an experiment. However, if the carbon chain is severed during the experiment, then correct interpretation of experimental results may require the presence of the label at only one position in the compound.

The fraction of the total activity of a sample which is contributed by the desired nuclide is described as the radionuclidic purity (or radioisotopic purity) of the sample. Usually, radioactivity contributed by the daughter of the desired nuclide is not considered in the determination of radionuclidic purity; for example, ^{90}Y is not considered an impurity in a sample containing its parent ^{90}Sr. Nevertheless, the contribution of the daughter nuclide to the total activity of a sample should be recognized. For example, in a sample of ^{95}Zr in equilibrium with its daughter ^{95}Nb, two-thirds of the total activity of the sample is contributed by the daughter nuclide.

Radionuclidic impurities in a sample usually result from activation of impurities in the target subjected to neutron or charged-particle bombardment. Radionuclide impurities may not be troublesome if counting techniques are employed which discriminate against the radiation from the impurities. For example, radionuclidic impurities often are unimportant in a sample measured by a differential γ-counting technique, whereas they may interfere severly if the radiation is detected by a non-discriminatory method such as photographic film or a GM counter. In a radioactive sample used for an *in vivo* diagnostic procedure in humans, a long-lived radionuclidic impurity may significantly increase the amount of radiation delivered to persons subjected to the procedure. For example, the presence of ^{203}Hg ($T_{1/2} = 47$ days) should be minimized in chlormerodrin labeled with ^{197}Hg ($T_{1/2} = 65$ hr) and used for kidney scanning.

The chemical purity of a sample indicates the proportion of the sample which is in the desired chemical form, irrespective of the substitution of radioactive atoms for stable atoms in the sample. Chemical impurities accumulated as a product is synthesized usually are removed during purification of the product. After purification, however, chemical impurities may reaccumulate as the product is stored before use. Chemical impurities such as lead, arsenic, and tellurium are particularly undesirable in a sample intended for medical use.

A preparation is carrier-free if a particular element in the preparation is represented only by radioactive atoms. The specific activity is greater for a carrier-free preparation than for a similar preparation containing a stable carrier.

Analysis of Labeled Compounds

Analysis of the quality of a radioactive product is accompanied by a number of difficulties, including diminishing sample activity, personal radiation exposure, and the limited quantity of sample usually available for analysis. Consequently, analysis frequently requires special techniques and microchemical procedures.

Inorganic Compounds. Inorganic samples of high specific activity and small mass often are analyzed by techniques such as emission spectrography, colorimetric absorptometry, polarography, and colorimetric spot tests. Emission spectrography is particularly useful, because this analytic technique furnishes a record of essentially all elements in a sample. To prevent the spread of radioactive contamination, the electrodes of the spectrograph must be enclosed within a shield during analysis of the sample.

Radionuclidic impurities in a sample often may be measured by differential pulse height analysis (Chapters 12 and 15) and absorption measurements (Chapter 3) of the radiation emitted by the sample. Measurement of the half-life for radioactive decay of the sample also may reveal radionuclidic impurities in the sample. These procedures are useful with both inorganic and organic compounds.

Organic Compounds. Techniques used most frequently for analysis of the radiochemical purity of a labeled organic compound include reverse isotope dilution analysis and various procedures for chromatographic separation. Reverse isotope dilution analysis is used often to determine the purity of a radioactive sample and the degree of contamination of the sample by specific radiochemical impurities. In this procedure, a small fraction of a radioactive sample is mixed thoroughly with a large excess of the nonradioactive counterpart of the sample. A small fraction of the mixture is removed, and the remaining mixture is purified by an appropriate method. The specific activity of the purified mixture is compared to that of the sample set aside before purification. If the original sample is pure, then the specific activities will be identical for the two samples of the mixture. If the original sample is only 90% radiochemically pure, then the specific activity of the purified mixture will be 90% of the specific activity of the sample set aside before purification.

Results of reverse isotope dilution analysis may be misleading if an isotope effect is present; if the labeled compound undergoes a chemical transformation during purification; if the nonradioactive counterpart of the sample is impure; or if purification is incomplete.

Paper and thin-layer chromatography are used widely for determining the radiochemical purity of labeled organic compounds. To separate and identify all the impurities in a radioactive sample, two or more solvents may be required. The amounts of various impurities in the sample may be measured by scanning the paper chromatogram or thin-layer plate with a suitable radiation detector.

Chromatographic analysis of radiochemical purity is rapid and sensitive, but is subject to a number of errors in technique and interpretation. For example, reactions of a labeled compound with the solvent or supporting medium sometimes furnish products which may be interpreted incorrectly as impurities in the original sample. Differences in evaporation rates of the labeled compound and impurities from the supporting media can influence the quantitative determination of radiochemical purity.

The chemical purity of labeled organic compounds usually is determined by gas chromatography or by ultraviolet, visible, or infrared absorption techniques. Gas chromatography is rapid and sensitive but is restricted to gases, volatile compounds, and compounds with reproducible patterns of pyrolysis. Of the various absorption techniques, infrared absorption is used most frequently.

A few examples of the detection and analysis of impurities in labeled compounds are described in the booklet *Purity and Analysis of Labeled Compounds* [11].

Instability of Labeled Compounds

The decomposition of an organic compound during storage has been recognized for many years as a potential source of error when the compound is removed from storage and used. The probability that a compound has changed during storage may be increased if the compound is radio-active, because the radiation emitted by the radioactive label may accelerate decomposition. The degree of change depends upon the conditions and duration of storage, and often is reduced by storing labeled compounds under conditions such as reduced temperature, *in vacuo*, or as a freeze-dried solid. Optimal storage conditions for a variety of labeled organic compounds are described by Bayly and Evans[12].

In general, radiation induced decomposition of a labeled compound increases with:

(1) The fraction of the energy emitted by the sample which is absorbed by molecules of the sample. Only a small fraction of the γ-ray energy outside the sample, and radiation-induced decomposition does not contribute significantly to sample degradation. Similarly, high-energy β particles such as those from ^{32}P expend much of their energy outside the sample, and radiation-induced decomposition usually is less severe for compounds labeled with ^{32}P than for compounds labeled with isotopes such as ^{14}C or ^{3}H which emit β particles of lower energy.

(2) The specific activity of the sample. In general, decomposition increases with the specific activity of the sample, because more energy is released and absorbed within a smaller mass of material.

(3) The time of storage of the sample. The extent of radiation-induced decomposition depends not only upon the storage time, but also upon the half-life of the radioactive label and the susceptibility of the labeled compound to radiation-induced degradation.

Various mechanisms for radiation-induced decomposition of labeled compounds have been identified[13, 14] and are outlined in Table 6-1.

Primary Internal Radiation Effect.   When a β-emitting atom undergoes radioactive decay, it is transformed into an atom of another element. The transformed atom may be retained by the molecule or it may be rejected. In either case, a chemical impurity is produced. A radiochemical impurity is produced by this process only if the original molecule contained at least two radioactive atoms. The percent radiochemical impurity introduced into a uniformly labeled sample by the primary internal radiation effect may be estimated with the expression:

$$\% \text{ radiochemical impurity} = 100\left[1 - \exp-\left(\left(\frac{n-1}{n}S_0 t\right)6.14 \times 10^{-14}\right)\right]^*$$

where n is the number of atoms of the element of interest per molecule,

*$\exp(-x)$ means e^{-x}

TABLE 6-1 *Mechanisms for Decomposition of Labeled Organic Compounds*

Mechanism for decomposition	Cause	Method for control
Primary internal radiation effect	Radioactive decay	None, for a particular specific activity
Primary external radiation effect	Interaction of radiation directly with molecules of compound	Dispersal of labeled molecules
Secondary radiation effect	Interaction of reactive intermediates with molecules of compound	Dispersal of labeled molecules; reduced temperature; scavenging of free radicals
Chemical effects	Chemical reaction unrelated to radiation	Reduced temperature; removal of harmful agents

S_0 is the initial specific activity of the sample in curies per mole, and t is the time in seconds since preparation of the sample. For compounds labeled with radioactive atoms of relatively short half-lives, the term t in the equation should be replaced by $(1 - e^{-\lambda t})/\lambda$, where λ is the decay constant for the radioactive isotope. In general, decomposition of a sample by the primary internal radiation effect increases in severity with increasing size of the labeled molecules.

Example 6-1

Estimate the percent radiochemical impurity after 2 years of storage for a sample of sucrose labeled uniformly with ^{14}C at an isotopic abundance of 20%.

A molecule of sucrose contains 12 atoms of carbon. If 20% of the carbon atoms are ^{14}C, then the specific activity of the sample in curies per mole is

$$\text{specific activity (Ci/mole)} = \frac{\lambda N}{3.7 \times 10^{10}}$$

$$= \frac{[(12)(0.2) \text{ atoms/molecule}]}{3.7 \times 10^{10} \text{ dis/sec-Ci}}$$

$$\times [6.02 \times 10^{23} \text{ molecules/mole}][0.693/(5600 \text{ yr})$$

$$\times (365 \text{ day/yr})(24 \text{ hr/day})(3600 \text{ sec/hr})]$$

$$= 1.53 \times 10^2 \text{ Ci/mole}$$

The percent radiochemical impurity is

$$\text{radiochemical impurity}, \% = 100\left[1 - \exp-\left(\left(\frac{n-1}{n}S_0 t\right)6.14 \times 10^{-14}\right)\right]$$

$$= 100\left[1 - \exp-\left(\left(\frac{12-1}{12}\right)(1.53 \times 10^2)(2)(365)\right.\right.$$

$$\left.\left. \times (24)(3600)(6.14 \times 10^{-14})\right)\right]$$

$$= 100\left[1 - \exp-0.00054\right]$$

$$= 0.06$$

Primary External Radiation Effect. Molecules of a labeled sample may be degraded during interactions of radiation emitted by adjacent molecules in the sample. This cause of radiation-induced decomposition is known as the primary external radiation effect. The extent of decomposition of a labeled compound by this effect is influenced by:

(1) Average energy $\bar{E}$ in electron volts which is released during decay of the radioactive label
(2) Fraction f of the released energy which is absorbed by the compound
(3) Susceptibility of the compound to radiation-induced changes; the susceptibility is described by the G value for the compound, defined as the number of molecules irreversibly changed per 100 eV absorbed
(4) Initial specific activity S_0 in curies per mole
(5) Time t of storage in seconds

The percent decomposition due to the primary external radiation effect may be estimated with the expression:

$$\text{decomposition}, \% = 100[1 - \exp-((\bar{E}fGS_0 t)6.14 \times 10^{-16})]$$

For compounds labeled with radioactive atoms of relatively short half life, the term t in the equation should be replaced by $(1 - e^{-\lambda t})/\lambda$.

Example 6-2
Estimate the percent decomposition over a storage period of 22 weeks for a sample of tritium-labeled thymidine with a specific activity of 14,000 mCi/mmole. The G value is 0.05, the average energy $\bar{E}$ is 0.0055 MeV for ^{3}H, and the fraction f of the energy absorbed is 1.

$$\text{decomposition}, \% = 100[1 - \exp - ((\bar{E}fGS_0t)6.14 \times 10^{-16})]$$

$$= 100[1 - \exp - ((5.5 \times 10^3)(1)(0.05)(1.4 \times 10^4)$$

$$\times (22)(7)(3600)(6.14 \times 10^{-16}))]$$

$$= 100[1 - \exp - 0.032]$$

$$= 3.0$$

Secondary Radiation Effect. Decomposition of a labeled compound may be initiated by reactive intermediates such as free radicals which are released as radiation interacts with the compound or with the solvent, impurities or container in the vicinity of the compound. Often, secondary radiation effects are the main cause for decomposition of a labeled sample during storage. The extent of decomposition caused by secondary radiation effects varies greatly with the storage conditions of a compound. For example, compounds stored in aqueous solution are exposed to a variety of reactive intermediates (such as hydrogen and hydroxyl free radicals, peroxides and hydrated electrons) produced during interactions of radiation with water. To reduce the influence of these reactive intermediates, labeled compounds in aqueous solution often are stored at reduced temperature. Also, these samples may contain "scavenger impurities" (for example, sodium formate, benzyl alcohol, ethanol, and cysteamine) which deactivate reactive intermediates and reduce their influence on the labeled compound. When radioactive solutions containing scavenger impurities are used in tracer studies, the possible influence of the impurities on experimental results should be considered.

Chemical Effects. During storage, all organic compounds decompose to some extent by mechanisms such as oxidation, hydrolysis, and biologic reactions. These mechanisms sometimes are accelerated in labeled compounds, particularly when the compounds are supplied in a very small volume.

Methods for Reducing Decomposition

(1) The time between purification and use of a compound should be as short as possible, because radiation-induced decomposition increases with storage time.

(2) The specific activity of a sample should be no greater than that required for the experiment envisioned for the sample, because sample degradation increases with specific activity. Frequently,

radioactive samples are diluted intentionally with a foreign substance which absorbs the radiation energy without producing reactive intermediates. The foreign substance usually is a solvent such
as benzene, sometimes mixed with a secondary solvent such as
dioxane or alcohol. Water and other hydroxylic solvents should
be used with caution, because they may enhance the contribution
of secondary radiation effects.

(3) One method to reduce sample decomposition is to distribute the
sample over a large area, permitting the radiation to escape from the
sample with as few interactions as possible. This method is useful
for many compounds labeled with isotopes such as ^{14}C, ^{35}S, and
^{32}P; it is not useful for compounds labeled with tritium, because the
low-energy β particles from ^{3}H are absorbed over too short a range.

(4) Secondary radiation effects and chemical effects may be suppressed
by storing a labeled sample at reduced temperature. When a sample
is frozen, the mobility of reactive intermediates is severely limited.
However, the intermediates may remain reactive and interact with
the labeled compound when the sample is thawed. Also, even though
a sample has been solidified, energy absorbed by solvent molecules may be transferred to molecules of the labeled compound and
produce degradation. Consequently, most liquid labeled samples
should be stored at a temperature just above the freezing point of
the solvent.

(5) Compounds stored as solids should be as free of moisture as possible, because reactive intermediates produced in the water contribute to sample degradation. Consequently, solid samples such
as freeze-dried compounds usually are stored in evacuated ampules.

(6) Free radical scavengers in a labeled sample often reduce the contribution of reactive intermediates to decomposition of the sample.
However, many of the conventional scavenging compounds are
undesirable because they increase sample degradation caused by
chemical effects, and also because they sometimes interfere with
subsequent applications of the labeled compound. To reduce
secondary radiation effects, a few substances which are relatively
inocuous have been added to labeled samples. For example, the
decomposition of labeled amino acids of high specific activity has
been reduced by a factor of six by the addition of 2% ethanol[1].
Other substances which have proved effective include benzyl
alcohol, cysteamine, and sodium formate.

(7) If a compound may be labeled and used satisfactorily with two or
more radioactive isotopes of the same element (for example, ^{125}I
or ^{131}I, ^{57}Co or ^{58}Co), then radiation-induced decomposition can be

reduced by choosing the isotope which produces the least decomposition. For example, triiodothyronine and insulin undergo less decomposition if labeled with ^{125}I rather than ^{131}I.

(8) Labeled compounds always should be stored in clean containers to reduce chemical and secondary radiation effects. The presence of a bacteriostat also may be desirable.

(9) Compounds which are sensitive to light should be stored in the dark.

Radiation-induced decomposition of a labeled compound usually is negligible during the course of an actual experiment, provided that the experiment is completed in a few hours or days. Radioactive compounds stored as solids sometimes assume a different color from their non-radioactive counterparts, because ions and reactive intermediates produced by the radiation and trapped in the crystal lattice exhibit intense optical absorption. Usually, the amount of impurity causing the change in appearance is negligible, and the sample resumes its normal color in solution.

Tables listing storage conditions and percent decomposition for a variety of labeled compounds have been prepared by Bayly and Evans [12].

PROBLEMS

1. Estimate the percent radiochemical impurity caused by the primary internal radiation effect over a storage period of 1 year for a sample of glucose labeled uniformly with ^{14}C at a ratio of $1:2$ for $^{14}C:^{12}C$ atoms.

2. Estimate the percent decomposition caused by the primary external radiation effect for a sample of DL-leucine-4,5-T stored for 30 weeks. The specific activity is 5400 mCi/mmole and the G value is 0.9.

REFERENCES

[1] Wilson, B. (ed.). 1966. The radiochemical manual. The Radiochemical Centre, Amersham, England.

[2] Wilzbach, K. 1957. Tritium labeling by exposure of organic compounds to tritium gas, J. Am. Chem. Soc. 79:1013.

[3] Isbell, H., H. Frush, and J. Moyer. 1960. Tritium-labeled compounds IV, D-Glucose-G-T, D-Xylose-5-T, and D-Mannitol-1-T, J. Res. N.B.S., 64A:359.

[4] Calvin, M., et al. 1949. Isotopic carbon. Techniques in its measurement and chemical manipulation, John Wiley & Sons, New York.

[5] Whisman, M., and B. Eccleston, 1962. Gas-exposure labeling of organics with tritium. Nucleonics 20(6):98.

[6] Wang, C., and D. Willis. 1965. Radiotracer methodology in biological science. Prentice Hall, Englewood Cliffs, N.J.

[7] Catch, J. 1961. Carbon-14 compounds. Butterworths, London.

[8] Kamen, M. 1957. Isotopic tracers in biology: an introduction to tracer methodology, 3rd ed. Academic Press, New York.

[9] Stepka, W., and P. Larson. 1965. Isotopes in experimental pharmacology, L. Roth, (ed.). University of Chicago Press, 7.

[10] Friedlander, G., J. Kennedy, and J. Miller. 1964. Nuclear and radio-chemistry, 2nd ed. John Wiley & Sons, New York.

[11] Catch, J. Purity and analysis of labeled compounds, The Radiochemical Centre, Amersham, England.

[12] Bayly, R., and E. Evans. Storage and stability of compounds labeled with radioisotopes, parts I and II. Radiochemical Division, Nuclear-Chicago Corp., Des Plaines, Ill.

[13] Bayly, R., and H. Weigel. 1960. Self-decomposition of compounds labeled with radioactive isotopes. Nature 188:384.

[14] Geller, L., and N. Silberman. 1970. Stability of labeled components after storage. *In* E. Bransome, Jr. (ed.), The current status of liquid scintillation counting. Grune & Stratton, New York.

7

Radioactive Pharmaceuticals

Radioactive pharmaceuticals are used widely for evaluating the structural and functional integrity of biologic systems and for treating certain diseases. The suitability of a particular pharmaceutical for a specific diagnostic or therapeutic procedure depends upon the characteristics of the pharmaceutical and its radioactive label. These characteristics include:

(1) Biologic behavior of the pharmaceutical
(2) Rate of decay of the label
(3) Radiations emitted by the label and their detectability with the apparatus available
(4) Contributions of the radiations to the total radiation dose absorbed by the patient
(5) Cost of producing the radioactive isotope and the labeled pharmaceutical

Listed in Table 7-1 are labeled pharmaceuticals considered "well-established" for human use. Although many other radioactive pharmaceuticals have been prepared, their use in humans currently is considered "investigational" rather than well-established.

Choice of Radioactive Nuclide

Various properties of a radioactive nuclide determine its usefulness as a label for a pharmaceutical. Among these characteristics are:

(1) Types, energies, and fractional emissions of radiation from the

***TABLE* 7-1**   *Well-Established Radioactive Pharmaceuticals*

Nuclide	Chemical Form	Use
^{32}P	Soluble phosphate	Treatment of polycythemia vera; bone metastases; leukemia
	Colloidal chromic phosphate	Intracavitary treatment of pleural and peritoneal effusions; ascites; interstitial treatment of cancer
^{42}K	Chloride	Potassium space studies
^{51}Cr	Chromate and chloride	Spleen scans (labeled RBC's); red cell tagging and survival studies; plasma and blood volume determinations
	Human serum albumin	Diagnosis of GI protein loss
^{57}Co, ^{58}Co, and ^{60}Co	Labeled vitamin B_{12}	Intestinal absorption studies
^{59}Fe	Chloride, sulfate, citrate	Iron turnover studies
^{60}Co	Wire alloy	Intracavitary and interstitial treatment of cancer
^{85}Sr	Nitrate and chloride	Bone scans in cases of diagnosed cancer
^{99m}Tc	Pertechnetate	Brain scans; thyroid scans; blood pool scans; placental investigation
	DTPA	Kidney scans
	Sulfur colloid	Liver and spleen scans
^{125}I and ^{131}I	Iodide	Diagnosis of thyroid function and thyroid scans
^{125}I and ^{131}I	Human serum albumin	Blood and blood plasma volume determinations; cisternography and ventriculography; cardiac scans; brain tumor localization
	Labeled renal function compounds	Kidney function studies
	Labeled fats and fatty acids	Fat absorption studies
	Iodide	Treatment of hyperthyroidism; cardiac dysfunction; thyroid carcinoma
	Rose bengal	Liver function and liver scans
^{131}I	Sodium iodipamide	Cardiac scans
	Human serum albumin microaggregates	Liver scans
	Human serum albumin macroaggregates	Lung scans

TABLE 7-1 (cont.)

Nuclide	Chemical Form	Use
	Labeled renal function compounds	Kidney scans
^{133}Xe	Gas	Cerebral blood flow; pulmonary function; muscle blood flow studies
^{137}Cs	Microspheres in steel needles	Intracavitary treatment of cancer
^{197}Hg and ^{203}Hg	Chlormerodrin	Brain scans
^{197}Hg	Chlormerodrin	Kidney scans
^{198}Au	Colloidal suspension	Liver scans; intracavitary treatment of pleural and peritoneal effusions; ascites; interstitial treatment of cancer
	Seeds	Interstitial treatment of cancer

nuclide. For example, γ-ray photons of 100–400 keV are desirable from a nuclide used in a scanning procedure, because photons in this energy range escape from the body and interact in the detector with reasonable frequency. Particulate radiations (for example, β particles, conversion and Auger electrons) are not desirable from a nuclide used in a scanning procedure, because these radiations increase the radiation dose absorbed by the patient without providing additional diagnostic information.

(2) Half-life for radioactive decay of the nuclide. Short-lived nuclides (for example, half-life < 1 hr) are difficult to manipulate chemically, cannot be stored very long before use, and may furnish results which are difficult to interpret because the activity decreases significantly over the course of the procedure. Long-lived nuclides (for example, half-life > 1 yr) may deliver excessive radiation doses to patients.

(3) Chemical form of the nuclide. The chemical form should lend itself to synthesis of a labeled pharmaceutical with the desired biologic properties.

(4) Contamination of the nuclide with impurities which are difficult to separate chemically. If the impurities are radioactive, then the radiation dose absorbed by the patient is increased.

Consideration of these characteristics has influenced some investigators to choose ^{125}I rather than ^{131}I for certain clinical studies, because ^{125}I decays by electron capture whereas ^{131}I decays by negatron emission. Consequently, the contribution of particulate radiations to the radiation dose to the patient is reduced with [125]I. Additionally, pharmaceuticals labeled with ^{125}I decompose less rapidly than those containing [131]I. The low-energy photons emitted by [125]I (~ 30 keV) are attenuated severely in tissue, and compounds labeled with this nuclide are unacceptable as scanning agents for organs deep within the body. For procedures which utilize iodine but do not require a complex pharmaceutical (for example, thyroid studies with iodide ion), the short-lived nuclide [123]I ($T_{1/2} = 13.1$ hr) may be preferable to either ^{125}I or [131]I. This isotope of iodine is produced in a cyclotron by the reaction [$^{121}Sb(^{3}He,n)^{123}I$ or $^{123}Te(p,n)^{123}I$].

The 140 keV γ-ray photons from ^{99m}Tc usually escape from the body without alteration, are easily collimated before they reach the detector, and are absorbed with reasonable efficiency in scintillation detectors. Consequently, many investigators consider ^{99m}Tc to be almost ideal for many scanning procedures. This nuclide is obtained by eluting a generator containing its longer lived parent (^{99}Mo). A variety of pharmaceuticals, including a colloid and albumin, can be labeled conveniently with ^{99m}Tc. During decay of ^{99m}Tc, no particulate radiation is emitted from the nucleus.

Choice of Pharmaceutical

The physical and chemical properties of a pharmaceutical determine its accumulation and retention in a particular organ or region within the body. Some of the processes by which labeled pharmaceuticals are accumulated and retained are summarized in Table 7-2. For a particular procedure, a radioactive pharmaceutical should be chosen which is retained in an organ or region of interest for a time adequate to perform the desired measurements. After the measurements have been completed, the pharmaceutical should be excreted rapidly and completely from the body, particularly if the half-life for radioactive decay is long.

A few compounds of biologic importance can be labeled isotopically and are used widely for studies of organ function. Examples of useful isotopically labeled compounds include thyroid hormone labeled with radioactive iodine and cyanocobalamin labeled with radioactive cobalt. Most compounds of biologic importance contain only carbon, hydrogen, oxygen, nitrogen, phosphorus, and sulfur, and γ-emitting isotopes of these elements are not available with half-lives convenient for clinical procedures. However, many compounds can be labeled nonisotopically to furnish radioactive pharmaceuticals which may be suitable for certain applications. The suitability of these pharmaceuticals depends primarily

TABLE 7-2 *Processes of Accumulation and Retention of Pharmaceuticals*[a]

Active transport	Capillary blockage
Thyroid scanning with iodide	Lung scanning with macroaggregated particles
Renal scanning with chlormerodrin	Renal scanning with particles
Liver scanning with rose bengal	Brain scanning with particles
Phagocytosis	**Simple or exchange diffusion**
Liver scanning with colloidal particles	Bone scanning with ^{18}F, ^{85}Sr, ^{87m}Sr, ^{47}Ca
Spleen scanning with colloidal particles	Brain scanning
Bone marrow scanning with colloidal particles	
Cell sequestration	**Compartmental localization**
Spleen scanning with damaged RBC's	Cardiac scanning for pericardial effusion
	Mediastinal scanning for aneurysms
Spleen scanning with mercuri-hydroxypropane	Placental scanning

[a]From H. Wagner 1966. Nuclear medicine: present and future. Radiology 86: 601.

upon the stability of the radioactive label and the degree to which the labeled compound mimics the behavior of its stable counterpart. A radioactive pharmaceutical designed for a particular application should not be used indiscriminately for another purpose. For example, human serum albumin labeled nonisotopically with ^{131}I is suitable for studies of the circulatory system; however, this compound may not be acceptable for metabolic studies[1].

Some compounds containing sulfur may be labeled with radioactive ^{75}Se in place of the sulfur atoms. In certain situations, these radioactive analogues behave similarly to their sulfur-containing counterparts. For example, L-selenomethionine labeled with ^{75}Se is incorporated into the pancreas in place of methionine and is used for studies of pancreas function. Iodine-containing compounds designed as contrast agents for diagnostic radiology may be labeled with radioactive iodine and used as radioactive pharmaceuticals. For example, Urokon, Diodrast, and Hippuran* labeled with radioactive iodine have been used for the study

*Urokon (Mallinckrodt Chemical Works), Diodrast (Winthrop Laboratories), Hippuran (Mallinckrodt Chemical Works).

of the clearance of substances from the kidney. Glomerular filtration rates have been studied using Hypaque† labeled with radioactive iodine. Other pharmaceuticals such as cyanocobalamin labeled with ^{57}Co and allyl inulin labeled with ^{125}I have also been used for this study.

Radionuclidic Purity

Radionuclidic impurities are undesirable in a labeled pharmaceutical, particularly if the impurities emit particulate radiation or have a relatively long half-life for radioactive decay. Consequently, the radionuclidic purity of a pharmaceutical should be verified whenever radionuclidic impurities may be present. Determination of the radionuclidic purity may require no more than analysis of the radiations emitted by the sample by a technique such as γ-ray spectrometry. More often, the determination involves chemical separation of an analytic sample, sometimes with the addition of a carrier, followed by physical analysis of the individual fractions. An alternate procedure is to allow the nuclide of interest to decay, and then to examine the residue for long-lived impurities. The retrospective nature of this method is a disadvantage.

Radiochemical Purity

The presence of radiochemical impurities in a radioactive pharmaceutical is undesirable because the impurities may exhibit a biologic behavior different from that of the pharmaceutical. In a particular application of the pharmaceutical, impurities may furnish misleading results as well as deliver an unnecessary radiation dose to the patient. The radiochemical purity of a pharmaceutical should be verified by paper or thin-layer chromatography, or by an alternate method such as reverse isotope dilution analysis.

Sterility

Most radioactive pharmaceuticals are administered intravenously and must be sterile, free from pyrogens, and of an acceptable pH. Many pharmaceuticals can be sterilized in their final containers by autoclaving conditions (that is, heating for 15–20 min at a temperature of about 260°F and at a pressure of roughly 20 lb/in.^{2}). The major alternative to autoclaving for producing sterile preparations is the passage of the preparation through a bacterial filter, usually a membrane filter composed of cellulose esters. Subsequent bottling and dispensing of the preparation

---

†Hypaque (Bayer Products Company).

must be performed under aseptic conditions. A third approach to sterilization which has been used occasionally is exposure of a preparation to high
levels of γ-radiation. For example, ^{32}P-labeled diisopropyl fluorophos-
phate in propylene glycol solution may be sterilized with a dose of 2.5
Mrad delivered by γ-radiation [2].

Verification of the sterility of radiopharmaceutical preparations is
accompanied by two problems:

(1) The period of 7 days required for a standard sterility test. For
many labeled pharmaceuticals, this period of time is available only
retrospectively.
(2) Possible effects of irradiation on the results of the sterility test,
together with the difficulties of manipulating radioactive preparations.

Occasionally, a compound cannot be subjected to a complete sterility
test before use. In these situations, the test period may be shortened
to perhaps 48 hr, provided that all conditions for preparing and injecting
the pharamaceutical are aseptic, and that complete retrospective testing
of the pharmaceutical is performed.

Often, the addition of a bacteriostat to a pharmaceutical preparation
helps to maintain the sterility of the preparation. The bacteriostat must
retain its potency and not interfere with the use of the preparation. Many
bacteriostats decompose during irradiation, and their potency decreases
with time. Even benzyl alcohol in 0.9% solution, the bacteriostat added
most often to radioactive pharmaceuticals, is not always useful because:

(1) It is a vasodilator and, therefore, cannot be added to a pharmaceutical such as ^{133}Xe in saline solution which is used for measurement of regional blood flow.
(2) It undergoes radiation-induced decomposition and produces a
precipitate (presumably benzoic acid) in certain solutions of high
specific activity.

For sterility testing, the *United States Pharmacopoeia* [3] specifies the
use of one of two media, fluid thioglycollate and fluid Sabouraud medium.
After inoculation, the thioglycollate tubes are incubated at 30–32°C for
7 days or more, and the tubes containing Sabouraud medium are incubated
at 22–25°C for at least 10 days.

Pyrogenicity

Most febrile reactions associated with pharmaceuticals are caused by
pyrogens, heat-stable substances produced by bacteria, viruses, yeasts, or

molds present in the pharmaceutical preparation. Symptoms of pyrogenicity include monophasic or biphasic fever, chills, malaise, mild to moderate pain in joints, leukopenia, and less definite signs such as apprehension, pallor, and substernal oppression. Bacterial products termed *endotoxins* are the major cause of pyrogenic reactions, and living or dead organisms need not be present in a preparation for the preparation to be pyrogenic. Consequently, sterilization does not inactivate pyrogens, and sterility is not necessarily indicative of the lack of pyrogens. Pyrogens are inactivated by autoclaving at 160°C for 10 hr or more; however, adequate precautions during preparation are the most reliable assurance against the presence of pyrogens in a product.

The *United States Pharmacopoeia*[3] outlines a procedure for testing a pharmaceutical for the presence of pyrogens. The test involves the use of three healthy mature rabbits, weighing not less than 1.5 kg and with a control temperature not above 39.8°C. Within 30 min after measurement of the control temperature of each animal, an appropriate amount (usually 3–10 times the equivalent human dose on the basis of weight) of the pharmaceutical is injected into a marginal vein of an ear of each rabbit. The temperature of each animal is measured each hour for 3 hr. If none of the animals exhibits a temperature rise greater than 0.6°C, and if the sum of the temperature rises in all three animals does not exceed 1.4°C, then the pharmaceutical may be considered free of pyrogens. If these conditions are not satisfied, then the test must be repeated with five additional rabbits. If not more than three of the eight rabbits exhibit a temperature rise greater than 0.6°C, and if the sum of the eight rises in temperature does not exceed 3.7°C, then the pharmaceutical may be considered free of pyrogens.

Examples of Diagnostic Agents for Scanning

Comprehensive reviews have been published of the medical uses of radioactive pharmaceuticals[4–8]. A few of these uses are outlined here.

Thyroid.   Since the iodide ion is concentrated selectively in the thyroid gland, it is the preferred material for diagnostic studies of the thyroid. Although ^{131}I in ionic form is used widely for examination of the thyroid, ^{125}I provides thyroid scans of comparable resolution and delivers less radiation dose to the patient. The nuclide ^{123}I provides γ-ray photons of desirable energy (159 keV) and has a short half-life (13.1 hr); however, this isotope is produced by positive ion bombardment and is not yet widely available. The physiologic behavior of ^{99m}Tc in the form of the pertechnetate ion (TcO_4^-) parallels the behavior of I^-, and some thyroid studies have been conducted with this nuclide.

Bone. Calcium in ionic form (Ca^{2+}) is absorbed preferentially in hyperfunctioning areas of bone, and is not accumulated in necrotic areas. Although bone studies have been conducted with ^{47}Ca, this nuclide is not the best choice for bone scanning because it is expensive and emits high-energy γ-rays. The biologic behavior of strontium is similar to that of calcium, and ^{85}Sr and ^{87m}Sr are suitable nuclides for bone scanning. The positron-emitting isotopes ^{18}F and ^{68}Ga are gaining limited acceptance as pharmaceuticals for bone scanning. The metastable nuclide ^{99m}Tc in polyphosphate form also is being studied as a possible agent for bone scans.

Bone Marrow. Iron is removed rapidly from the bloodstream and deposited primarily in the bone marrow. The short-lived ($T_{1/2} = 8.3$ hr), positron-emitting nuclide ^{52}Fe has been used for bone marrow scanning. Colloidal particles of technetium sulfide (^{99m}Tc), iodinated albumin (^{131}I), and gadolinium hydroxycitrate (^{159}Gd) also have been used for bone marrow scanning.

Lungs. Macroaggregated albumin consists of particles $25-30\,\mu$ in diameter which are trapped temporarily in capillaries of the lung. Albumin particles labeled with ^{99m}Tc or ^{131}I furnish satisfactory lung scans. Particles of iron hydroxide and aluminum hydroxide labeled with ^{113m}In also have been used for lung scanning.

Placenta. Human serum albumin labeled with ^{131}I, ^{123}I or ^{99m}Tc, and ^{113m}In attached to transferrin, are used for visualization of the placental blood pool.

Kidneys. Mercury diuretics are concentrated in the renal parenchyma, and chlormerodrin labeled with ^{197}Hg or ^{203}Hg is used widely for kidney scanning. Although better resolution is obtainable with ^{203}Hg, this nuclide also delivers a greater radiation dose to the kidneys, and ^{197}Hg is preferred by most investigators. In some centers, ^{99m}Tc in reduced form and ^{68}Ga-polymethaphosphate-Mg-polymetaphosphate are being used for visualization of the kidneys.

Liver. Dyes such as rose bengal are absorbed selectively by the reticuloendothelial cells of the liver, and rose bengal labeled with ^{131}I is used as a pharmaceutical for liver scanning. The nuclides ^{198}Au, ^{99m}Tc, ^{131}I and ^{113m}In in colloidal form also are used for visualization of the liver, and ^{99m}Tc in the form of sulfur colloid is becoming recognized as the pharmaceutical of choice for this procedure.

Spleen. Damaged red blood cells concentrate rapidly in the spleen, and ^{51}Cr-labeled red blood cells which have been injured on purpose are used for spleen scanning. Colloidal particles of ^{131}I and ^{99m}Tc are used also for visualization of the spleen.

Pancreas. Some amino acids concentrate selectively in the pancreas, and an appropriate amino acid labeled with a suitable radioactive nuclide might be an acceptable agent for visualization of this organ. Although L-selenomethionine labeled with ^{75}Se is the most promising agent developed so far for scanning of the pancreas, results with this pharmaceutical often are unsatisfactory. In some cases, the parathyroid glands also can be visualized with this agent.

Heart Muscle. Cesium and rubidium salts injected intravenously are concentrated in muscle. Scanning of the myocardium and the demonstration of infarcts have been accomplished with ^{131}Cs, a nuclide which releases characteristic x-rays during decay by electron capture.

Brain. Concentration of a pharmaceutical in a brain tumor is related to passive diffusion resulting from a breakdown in the physiologic barrier between blood and brain. Agents used most commonly for brain scanning include ^{99m}Tc in pertechnetate form, ^{197}Hg- and ^{203}Hg-labeled chlormerodrin, ^{113m}In- and ^{68}Ga-labeled DTPA (diethylenetriamine pentaacetic acid), human serum albumin labeled with ^{131}I, and the noble gas nuclides ^{131m}Xe, ^{133}Xe, and ^{85}Kr.

Examples of Therapeutic Agents

Disorders of the thyroid gland often are treated with millicurie doses of ^{131}I in ionic form, and polycythemia vera and other blood dyscrasias may be treated with ^{32}P in the form of phosphate ion. Colloidal preparations of ^{198}Au and ^{32}P in the form of chromic phosphate, and ^{90}Y in the form of yttrium silicate, are used sparingly in the treatment of malignant serous effusions in the peritoneal and pleural cavities. A number of isotopes (for example, radium, ^{222}Rn, ^{198}Au, ^{182}Ta, ^{60}Co, ^{137}Cs, and ^{131}Cs) have been incorporated into sealed capsules for interstitial and intracavitary therapy.

REFERENCES

[1] Wilson, B. (ed.). 1966. The radiochemical manual. Amersham, England. 103.

[2] Charlton, J. 1966. Problems characteristic of radioactive pharmaceuticals. *In* G. Andrews, R. Kniseley, and H. Wagner. (ed.), Radioactive pharmaceuticals CONF-651111. Division of Technical Information, USAEC, Oak Ridge, Tenn.

[3] United States Pharmacopoeia, 17th rev. 1965. Mack Publishing Co., Easton, Pa.

[4] Silver, S. 1968. Radioactive nuclides in medicine and biology: medicine, 3rd ed. Lea and Febiger, Philadelphia.

[5] Wagner, H. (ed.). 1968. Principles of nuclear medicine, W. B. Saunders Co., Philadelphia.

[6] Hendee, W. 1970. Medical radiation physics. Year Book Medical Publishers, Chicago.

[7] McAfee, J., and G. Subramanian. 1966. Radioactive agents for the determination of body organs by external imaging devices; a review. Instr. Soc. Am. Trans. 5:349.

[8] Andrews, G., R. Kniseley, and H. Wagner (ed.). 1966. Radioactive pharmaceuticals CONF-651111. Division of Technical Information, USAEC, Oak Ridge, Tenn.

8

Gas-Filled Detectors

As ionizing radiation deposits energy in a gaseous medium, ion pairs are produced. The ion pairs may be collected by placing charged electrodes in the medium. If the electrodes collect only the ion pairs produced directly by the radiation, then the gas and charged electrodes constitute a type of radiation detector described as an ionization chamber. If additional ionization is produced and collected as the ion pairs released by the incident radiation migrate toward the collecting electrodes, then the detector is either a proportional counter or Geiger–Mueller tube.

Ionization Chambers

An ionization chamber designed to detect radiation from low-activity sources consists of parallel-plate or coaxial electrodes enclosing a gas-filled volume. In a parallel-plate chamber, the electrodes often are enclosed within a container comprised partly of a thin metal foil or mesh which permits the radiation to enter the chamber. A coaxial chamber consists of a central electrode in the form of a straight wire or wire loop which is charged positively with respect to the surrounding guard electrode or cylindrical case (Figure 8-1). The plastic insulator between the central electrode (anode) and the negatively charged guard electrode or case (cathode) is designed to minimize both the migration of charge through the insulator or along its surface (charge leakage) and the gradual penetration of charge into the body of the insulator (charge soak-in).

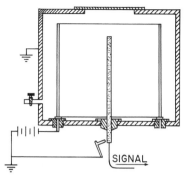

Figure 8-1 A simple coaxial ionization chamber, consisting of a central electrode (anode) surrounded by a negatively charged guard electrode (cathode) and cylindrical case.

Electroscopes. An ionization chamber used as an electroscope is charged by applying a voltage momentarily across the electrodes of the chamber. After the charging voltage is removed, the charge imbalance which remains between the chamber electrodes is indicated by the displacement of a quartz fiber attached to one electrode. When the chamber is exposed to radiation, the imbalance of charge is reduced. The rate of reduction in charge imbalance is proportional to the rate of entry of radiation into the chamber, and the total reduction in charge imbalance is proportional to the total exposure of the chamber to radiation. After exposure, the charge imbalance remaining between the electrodes is depicted by the residual displacement of the quartz fiber. The shadow of the quartz fiber is projected onto an illuminated scale, and the position of the shadow along the scale reflects the total amount of radiation to which the chamber has been exposed.

Pocket Chambers and Dosimeters. Electroscopes shaped as a fountain pen and used to monitor the exposure of persons to radiation are referred to as pocket chambers or pocket dosimeters (Figure 8-2). To charge a pocket chamber or dosimeter, a charger containing a battery or rectified power supply is used to apply voltage to the chamber electrodes. The voltage is adjusted until the shadow of the fiber falls at 0 on an illuminated scale. After exposure to radiation, a pocket chamber must be reinserted into the charger (termed the *charger-reader*) to determine the deflection of the fiber upscale. The pocket dosimeter contains its own scale, and the deflection of the fiber upscale may be determined by holding the dosimeter toward light.

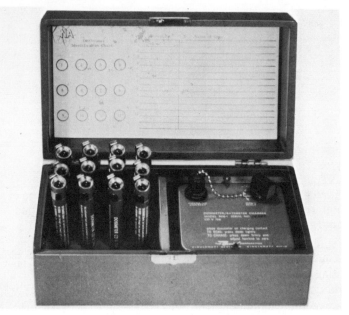

Figure 8-2 Pocket chambers and a charger-reader. (Courtesy of Nuclear Associates, Inc.)

Lauritsen Electroscope. This device is similar to the pocket dosimeter, except that it is bulkier (the chamber is 2–3 in. in diameter and about 3 in. long) and is used primarily for measurement of radiation from radioactive samples. In the instrument diagrammed in Figure 8-3, the deflection of the quartz fiber is viewed through the microscope at one end of the chamber.

The Landsverk electroscope or analysis unit is somewhat similar to the Lauritsen electroscope, except that the sample may be inserted and measured within the chamber.

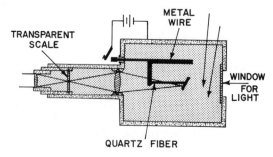

Figure 8-3 The Lauritsen electroscope.

Pulse-Type Ionization Chambers. As radiation interacts with the gas inside an ionization chamber, ion pairs (electrons and positive ions) are produced. The electrons migrate rapidly to the central electrode (anode), and the positive charge of this electrode is reduced as the electrons are collected. Usually, the electrons are collected within a microsecond or so after ionization of the gas. The heavier positively charged ions migrate more slowly toward the negative case (cathode) of the chamber. During their migration, they induce a negative charge on the cathode which partially masks the voltage reduction between the electrodes resulting from the collection of electrons. The induced charge is not removed until all positive ions have been neutralized at the cathode. Consequently, the total reduction in voltage between the electrodes is not attained until all positive ions within the chamber have been neutralized (Figure 8-4). The time required to collect the positive ions is a few hundred microseconds.

In the interest of fast response, most pulse-type ionization chambers utilize as a voltage pulse only the reduction in voltage furnished by the collection of electrons. The size V of the voltage pulse is

$$V = \frac{Q - q^+}{C}$$

where Q is the charge collected by the anode, C is the capacitance of the chamber plus associated electronics, and q^+ is a correction for the charge induced on the collecting electrode by the slowly moving positive ions.

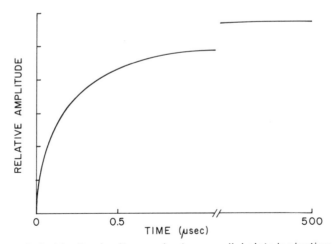

Figure 8-4 Idealized voltage pulse in a parallel plate ionization chamber. Electrons are collected within 0.5 μsec after their formation by impinging radiation, whereas 500 μsec or so are required for the collection of positive ions.

Example 8-1

What size voltage pulse is produced by total absorption of a 2.5 MeV α particle in a gas-filled ionization chamber with a capacitance of 20 picofarad (20 picofarad = 20×10^{-12} farad), if the induced charge reduces the effect of the charge collected by 20%?

The number of ion pairs produced by total absorption of a 2.5 MeV α particle in air is:

$$\text{number of ion pairs} = \frac{2.5 \times 10^6 \, \text{eV}}{33.7 \, \text{eV/IP}}$$
$$= 7.4 \times 10^4 \, \text{IP}$$

Consequently, 7.4×10^4 electrons and 7.4×10^4 positive ions are released. Since each electron possesses 1.6×10^{-19} coul of charge, the charge Q collected by the anode is

$$Q = (7.4 \times 10^4 \, \text{electrons}) \, (1.6 \times 10^{-19} \, \text{coul/electron})$$
$$= 11.8 \times 10^{-15} \, \text{coul}$$

The voltage pulse V produced is

$$V = \frac{Q}{C}$$
$$= \frac{(11.8 \times 10^{-15}) - (0.2) \, (11.8 \times 10^{-15}) \, \text{coul}}{20 \times 10^{-12} \, \text{farad}}$$
$$= 4.8 \times 10^{-4} \, \text{V}$$
$$= 0.48 \, \text{mV}$$

In a pulse-type chamber designed to produce voltage pulses by rapid collection of electrons, no interference with the migration of these electrons can be tolerated. Gases such as oxygen, water vapor, and the halogens have an affinity for electrons and should not be present in a pulse-type chamber. Gases such as helium, neon, argon, hydrogen, nitrogen, carbon dioxide, and methane combine with electrons only rarely to form negative ions, and these gases are used often as filling gases in pulse-type ionization chambers. The W quantities for these gases are included in Table 8-1.

Since the size of the voltage pulse produced in an ionization chamber depends partly upon the amount of charge induced upon the cathode, the site of production of ion pairs within the chamber influences the pulse size. This influence may be reduced by using a very small electrode as the anode or, more frequently, by surrounding the anode with a wire-mesh grid. The grid interferes minimally with the movement of electrons, and the voltage pulse produced as electrons are collected is not reduced by the charge induced by the positive ions. Gridded pulse-type ionization chambers have been used widely for the measurement of energy spectra

TABLE 8-1 *The Single Ionization Potential and* **W** *quantity for Selected Gases* [1]

Gas	Single ionization potential, eV	W quantity eV
H_2	15.6	36.9
N_2	15.7	34.9
O_2	12.5	31.3
He	24.6	41.3
Ne	21.6	35.9
Ar	15.8	26.3
Kr	14.0	24.4
Xe	12.1	22.1
CO_2	13.7	32.7
Air	—	33.7
CH_4	13.1	28.1

for charged particles, particularly α particles from radioactive samples[2]. Pulse-type chambers are used primarily for detection of heavy charged particles and neutrons, because the relatively large pulses furnished by these radiations can be amplified and measured conveniently. Other less densely ionizing radiations such as β particles are detected more conveniently with a current-type chamber.

Current-Type Ionization Chambers. Electrons collected at the anode of an ionization chamber constitute a direct current which may be amplified in an electrometer tube and measured with a conventional microammeter. This method is used often in portable monitoring instruments such as the "cutie pie" illustrated in Figure 8-5. In general, this approach is unsatisfactory for measurements of high accuracy, because instability and zero drift are introduced by problems associated with amplification of d-c signals and by spurious currents flowing across or through the chamber insulators.

Small currents from an ionization chamber may be measured accurately with a vibrating reed (dynamic capacitor) electrometer. The vibrating reed electrometer consists of a capacitor with one moving plate which oscillates at a frequency of 200–500 Hz. The signal from the ionization chamber is converted into an alternating voltage by the vibrating reed electrometer and amplified with an ac amplifier which is not subject to problems associated with dc amplifiers. The amplified ac signal is rectified to a dc signal and measured to within 0.05% precision by a conventional

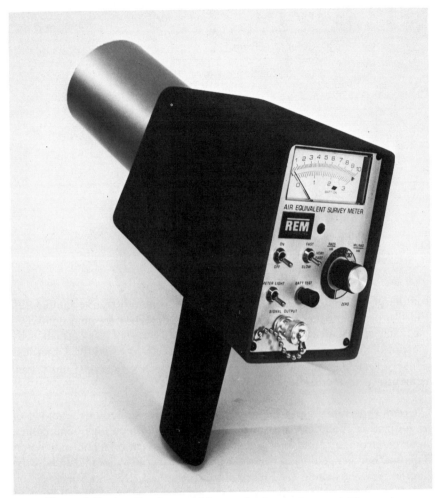

Figure 8-5 A cutie pie portable monitoring instrument. The scale of the meter is calibrated to read directly in units of mrads per hour and rads per hour in tissue, with a range switch which changes the sensitivity of the instrument by factors of 3. (Courtesy of REM, Inc.)

meter. The measurements are obtained by one of two methods, known as the voltage-drop method and the rate-of-charge method.

With the voltage-drop method, the current from the ionization chamber is used to charge a capacitor. The charge on the capacitor leaks through a precision resistance of 10^7–10^{12} ohm, producing a voltage drop across the resistance. This voltage is used by the vibrating reed electrometer to

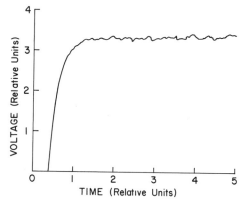

Figure 8-6   Typical strip chart recording of voltage as a function of time for an ionization chamber signal measured by the voltage-drop method.

produce an ac signal which is amplified, rectified, and measured (Figure 8-6). With the rate-of-charge method, the charge is not allowed to leak from the capacitor; instead, the increase in voltage across the capacitor is used as the input signal for the vibrating reed electrometer (Figure 8-7). The rate-of-charge method is more sensitive than the voltage-drop method, and may be used to measure ionization currents as low as 10^{-17} amp. With the rate-of-charge method, very low levels of radioactivity may be measured with an ionization chamber.

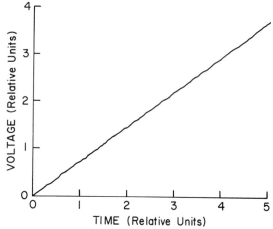

Figure 8-7   Typical strip chart recording of voltage as a function of time for an ionization chamber signal measured by the rate-of-charge method.

Example 8-2

A $0.005\,\mu\text{Ci}$ sample of $^{14}\text{CO}_2$ is contained within a CO_2-filled current-type ionization chamber. The ionization current is measured by the voltage-drop method, utilizing a precision resistance of 10^{11} ohm. If all of the energy of the β particles is deposited in the gas, what voltage is developed across the precision resistance?

$$(0.005\,\mu\text{Ci})\,(3.7 \times 10^4\text{ disintegrations/sec-}\mu\text{Ci})$$
$$= 185\text{ disintegrations/sec}$$

The average energy of β particles from ^{14}C is $0.05\text{ MeV} = 50 \times 10^3\text{ eV}$. The average energy dissipated per second in a counting volume is

$$(185\text{ disintegrations/sec})\,(50 \times 10^3\text{ eV/disintegration})$$
$$= 92.5 \times 10^5\text{ eV/sec}$$

The number of electrons released per second is

$$\frac{92.5 \times 10^5\text{ eV/sec}}{32.7\text{ eV/IP}} = 2.83 \times 10^5\text{ IP/sec}$$
$$= 2.83 \times 10^5\text{ electrons/sec}$$

where the W quantity is 32.7 eV/IP for CO_2 (Table 8-1).

The charge liberated per second is

$$(2.83 \times 10^5\text{ electrons/sec})\,(1.6 \times 10^{-19}\text{ coul/electron})$$
$$= 4.5 \times 10^{-14}\text{ coul/sec}$$
$$\text{where 1 coul/sec} = 1\text{ amp}$$
$$4.5 \times 10^{-14}\text{ coul/sec} = 4.5 \times 10^{-14}\text{ amp}$$

The voltage drop V across a resistance is the product of the resistance in ohms and the current in amperes

$$V = (4.5 \times 10^{-14}\text{ amp})\,(10^{11}\text{ ohm})$$
$$= 4.5 \times 10^{-3}\text{ V}$$
$$= 4.5\text{ mV}$$

Example 8-3

Repeat Example 8-2 for the case in which the rate-of-charge method of measuring ionization current is employed with a precision capacitance of 10 picofarad.

The rate of flow of charge is $4.5 \times 10^{-14}\text{ coul/sec}$. Since $V = Q/C$, where C is the capacitance, the rate of change of voltage $V_r = Q_r/C$, where Q_r is the rate of flow of charge.

$$V_r = \frac{4.5 \times 10^{-14}\text{ coul/sec}}{10 \times 10^{-12}\text{ farad}}$$
$$= 4.5 \times 10^{-3}\text{ V/sec}$$

The rate of change of voltage expressed as volts per minute is

$$V_r = (4.5 \times 10^{-3} \text{ V/sec}) (60 \text{ sec/min})$$
$$= 270 \times 10^{-3} \text{ V/min}$$
$$= 270 \text{ mV/min}$$

Various balance circuits have been devised for use with ionization chambers to circumvent problems of varying sensitivity caused by unstable operating conditions. With these circuits, a potential difference is applied to the ionization chamber to offset the voltage reduction produced by the ionizing radiation. After exposure of the chamber, the potential difference necessary to return the quartz fiber to 0 is measured and is indicative of the amount of ionization occurring in the chamber.

The current from an ionization chamber is plotted in Figure 8-8 as a function of voltage between the electrodes of the chamber. At low voltages, the electrons and positive ions are not separated quickly after their formation, and some of the ion pairs are lost by recombination. At higher voltages, the degree of recombination is reduced and the ionization current is greater. At voltages above the saturation voltage, essentially all ion pairs produced by the radiation are collected, and the ionization current is independent of voltage. The saturation voltage for an ionization chamber varies from 50 to 500 V and depends upon the design of the chamber, the shape and spacing of the electrodes, and the type and pressure of the gas in the chamber. In Figure 8-8, the region of the curve

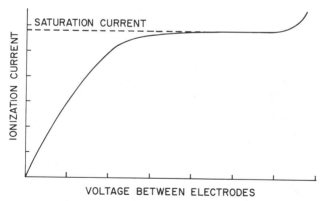

VOLTAGE BETWEEN ELECTRODES

Figure 8-8 Ionization current from an ionization chamber as a function of voltage between the electrodes. Recombination of ion pairs occurs at voltages below the saturation voltage. At high voltages, the signal is amplified by ionization produced as electrons released during radiation interactions are accelerated toward the anode.

above the saturation voltage is referred to as the *ionization chamber plateau*. The abrupt increase in ionization current at the high-voltage end of the plateau reflects the production of additional ion pairs as electrons released during radiation interactions are accelerated rapidly toward the anode. Ionization chambers are operated at a voltage below that required for production of additional ion pairs. At any particular voltage, the ionization current produced by a β-emitting sample is less than that produced by an α-emitting sample of equal activity, because β particles produce less ionization as they traverse the chamber.

Radiation from solid, liquid, and gaseous samples may be measured with an ionization chamber and vibrating reed electrometer. The activity of γ-emitting samples prepared for administration to animals or patients often is determined by placing the vial or syringe containing the sample into a well ionization chamber (Figure 8-9). Gaseous samples may be counted by filling an ionization chamber with the radioactive gas.

A neutron detector may be constructed by filling an ionization chamber with BF_3 gas or by coating the wall of an ionization chamber with lithium or boron. Ionization is produced in the chamber by interactions of α par-

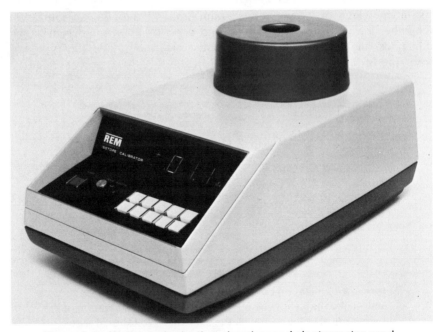

Figure 8-9 Well-type ionization chamber and electrometer used to measure the activity of radioactive samples. (Courtesy of REM, Inc.)

ticles and recoil nuclei liberated as neutrons interact with the lithium or boron.

Proportional Counters

The weak signal from an ionization chamber must be amplified greatly before it can be measured with conventional instruments. Electronic circuits used for signal amplification introduce electronic noise and instability which often distort or mask the signal. Signal distortion may be reduced by amplifying the signal within the counting volume of the chamber. This amplification is achieved by increasing the voltage between the chamber electrodes to a value above the ionization chamber plateau (Figure 8-8). Under the influence of the increased voltage, electrons released during radiation interactions are accelerated to a velocity great enough to produce additional ionization. The production of secondary ionization is controlled by using a coaxial chamber with an anode of a very small diameter, usually shaped as a small loop. With this configuration, almost all of the secondary ionization occurs in the immediate vicinity of the anode, and perhaps 10^6–10^7 electrons are collected for a much smaller number (10^3–10^5) of electrons liberated during radiation interactions. This process is referred to as *gas amplification*, and the amplification factor is the ratio of the total number of electrons collected to the number liberated directly by the incident radiation. The amplification factor varies from 10^2 to 10^4 for most proportional counters, and a signal of about 1 mV is furnished and requires only a small amount of external amplification.

The number of ion pairs collected in a gas-filled chamber is plotted in Figure 8-10 as a function of the voltage across the electrodes. In the proportional region, the amount of charge collected increases with the number of ion pairs produced initially by the radiation. Consequently, the size of the signal from a proportional chamber increases with the amount of ionization produced by the radiation traversing the chamber.

Example 8-4

An α particle produces 10^5 ion pairs in a proportional chamber with an amplification factor of 10^3. How many ion pairs are collected by the electrodes?

With an amplification factor of 10^3, 1000 ion pairs are collected for every ion pair liberated by the incident radiation. The total number of ion pairs collected is

$$10^5 \, IP \times 10^3 = 10^8 \, IP$$

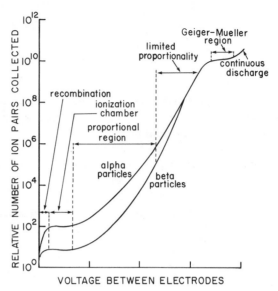

Figure 8-10 Total number of ion pairs collected in a gas-filled chamber as a function of the voltage across electrodes of the chamber.

Example 8-5

Repeat the calculation in Example 8-4 for a β particle which produces 10^3 IP within the chamber.

$$10^3 \text{ IP} \times 10^3 = 10^6 \text{ IP}$$

From the examples above, it is apparent that the signal produced in a proportional counter is considerably larger for an α particle than for a β particle. A "discriminator" may be used with the counter to reject all signals below a certain size. In this manner, signals produced by α particles may be recorded and those produced by β particles may be rejected. Consequently, only the α particles may be measured from a source which emits both α and β particles. By increasing the voltage across the chamber the signals produced by β particles may be amplified until they also are large enough to pass the discriminator. At the higher voltage, signals produced by either type of radiation are recorded. The counts due to β particles alone may be determined by subtracting the α counts measured at a lower voltage from the α-plus-β counts measured at a higher voltage.

A characteristic curve for a proportional counter is shown in Figure 8-11. If the counter is operated at a voltage along the α plateau, only α particles are detected. At these voltages, the count rate due to background

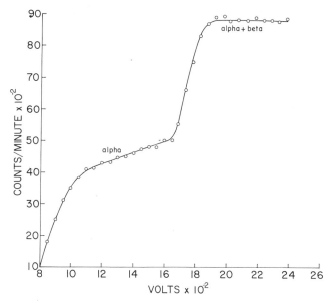

Figure 8-11 Characteristic curve for a proportional counter with P-10 gas.

radiation is very low. Both α and β particles are counted if the voltage is increased to the β plateau. For a well-designed proportional counter, the slope of the β plateau should not exceed 0.2%/100 V over a voltage range of several hundred volts. The slope of the plateau is computed by:

$$\text{slope} (\%/100 \text{ V}) = \frac{2[(\text{cpm})_2 - (\text{cpm})_1]10^4}{[(\text{cpm})_2 + (\text{cpm})_1](V_2 - V_1)}$$

where $(\text{cpm})_2$ is the count rate at voltage V_2 on the plateau, and $(\text{cpm})_1$ is the count rate at voltage V_1 on the plateau.

The voltage required for operation of a particular proportional counter on the α or β plateau depends upon the type of counting gas used and the pressure of the gas inside the chamber. If the counting gas is methane, for example, then the β plateau for most detectors is 3000–3500 V. The β plateau occurs between 2000 and 2500 V with P-10 gas, a mixture of 90% argon and 10% methane. Other gases used for proportional counting include 96% helium with 4% isobutane and 95% argon with 5% carbon dioxide.

Because of their low efficiency for γ-rays, proportional counters are used only rarely for the detection of γ-radiation from low-activity radioactive sources. Proportional counters are used widely, however, as detectors for high-intensity x-ray beams such as those encountered in x-ray spectrometry.

Because of the slow migration of positive ions to the cathode, a few hundred microseconds are required for the signal in a proportional counter to attain its maximum size. However, the signal reaches one-third to half of its maximum size after only about 5 μsec. Usually, a "clipping circuit" is used to obtain a pulse of adequate size within a time of 5–50 μsec for formation of the pulse. With this circuit, high count rates may be measured with a proportional counter, because interactions in the chamber are recorded as separate events if they are separated by just a few microseconds or more.

At voltages above the proportional region (Figure 8-10), α particles and other densely ionizing radiations ionize almost all the atoms of gas in the vicinity of the anode. If a chamber is operated at a voltage in this region, then the number of ion pairs collected is not proportional strictly to the ionization produced directly by the radiation. This voltage region is referred to as the *region of limited proportionality*, and is not utilized routinely in radioactivity measurements.

Geiger–Mueller Tubes

If the voltage between the electrodes of a gas-filled detector exceeds the region of limited proportionality, then the interaction of a charged particle or x- or γ-ray photon in the detector initiates an avalanche of ionization which results in almost complete ionization of the counting gas. Under these conditions, the number of ion pairs collected by the electrodes is independent of the amount of ionization produced directly by the impinging radiation. Consequently, the voltage pulses (1–10 V) emerging from the detector are of constant size and independent of the type of radiation initiating the signal. The range of voltage over which signals from the detector are independent of the type of radiation initiating the signal is referred to as the Geiger–Mueller or GM region, and detectors operated at voltages in this region are referred to as *GM detectors*. The amplification factor ranges from 10^6 to 10^8 for most GM detectors, and the voltage pulses from the detector require no additional amplification before they are counted.

In Figure 8-12, the number of voltage pulses (or counts) recorded per minute is plotted as a function of the voltage across the electrodes of a GM detector exposed to a radioactive source. No counts are recorded if the voltage is less than the starting voltage, because the pulses formed by the detector are too small to pass the discriminator, which usually is set at 250 mV. As the voltage is raised slightly above the starting voltage,

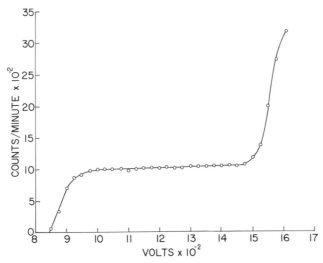

Figure 8-12 Characteristic curve for a GM detector quenched with an organic quench agent.

some of the pulses are transmitted by the discriminator and recorded. At the plateau threshold voltage, essentially all the pulses are transmitted to the scaler, and an increase in voltage beyond the threshold does not furnish a significant increase in count rate. Consequently, relatively inexpensive high-voltage supplies which are not exceptionally stable may be used with GM detectors. Usually, GM detectors are operated at a voltage about one-third of the way up the plateau. In Figure 8-12, for example, an operating voltage of 1150 V might be selected for the detector. The range of voltage encompassed by the plateau varies with the design of the GM detector and with the counting gas used.

If the voltage applied to a GM detector is raised beyond the plateau, atoms of counting gas may be ionized spontaneously. The voltage region beyond the plateau is termed the *region of spontaneous or continuous discharge*, because ionization within the detector may not have been initiated by radiation interactions. A GM detector can be damaged permanently by application of a voltage in the region of spontaneous discharge.

When radiation interacts in a GM detector, an avalanche of ionization is created along the entire length of the anode. Although the electrons are collected quickly, the residual positive ions require about 200 μsec to migrate away from the anode, and during this time the detector will not respond fully to additional radiation entering the counting volume. The curve in Figure 8-13 depicts the response of a GM detector as a function

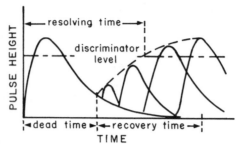

Figure 8-13 Formation of a second voltage pulse in a GM detector as a function of time after an initial ionizing event.

of time after an ionizing event. During the dead time, the detector is completely unresponsive to additional radiation. During the recovery time, an ionizing event produces a voltage pulse which is smaller than normal. The resolving time is the minimum time between an ionizing event and a second event which produces a voltage pulse large enough to pass the discriminator. For GM detectors, the resolving time usually is 200–300 μsec. A proportional counter often is preferred over a GM detector for the assay of a radioactive sample, because the resolving time of the proportional counter is lower and its plateau usually is flatter.

Positive ions which approach the cathode of a GM detector dislodge electrons from the wall of the detector. As these electrons combine with the positive ions, ultraviolet photons are released which are absorbed by the chamber wall and counting gas. The absorption of these photons may initiate the release of additional electrons which cause the chamber to remain discharged. Consequently, unless the emission of ultraviolet photons is "quenched," a GM detector will remain unresponsive after the first ionizing event. Self-perpetuating discharge of a GM detector is diagrammed in Figure 8-14 (left).

Various methods have been devised to quench the self-perpetuating discharge of GM detectors. With older GM counting systems, circuits were devised to quench the discharge externally by reducing the voltage across the detector momentarily after the formation of each signal. More recently, quenching has been achieved by adding a small concentration (about 0.1%) of a second gas to the counting gas. Gases used as internal quench agents include polyatomic organic gases (such as amyl acetate, ethyl alcohol, or ethyl formate) and halogens (Br_2 or Cl_2). The effect of a quench gas is depicted in Figure 8-14 (right). As the positive ions move toward the cathode, they collide with and transfer charge to molecules of the quench gas. The charged molecules of the quench gas migrate to the cathode and dislodge electrons from the chamber wall. As these electrons

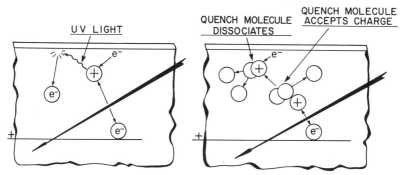

Figure 8-14 Left, self-perpetuating discharge of a GM detector caused by bombardment of the cathode or counting gas by ultraviolet photons released as positive ions are neutralized. Right, molecules of the quench gas accept the charge of positive ions and dissociate when neutralized near the cathode, thereby preventing the self-perpetuating discharge.

neutralize the charged molecules, energy is released which initiates dissociation of the quench molecules. The dissociation is irreversible for organic quench agents, and the useful life is 10^8–10^{10} pulses for a GM detector quenched with an organic quench agent. Halogen-quenched tubes have an infinite useful life, because the halogen atoms recombine after dissociation.

The counting gas used routinely in GM detectors is an inert gas such as argon, helium, or neon at a pressure of perhaps 100 mm Hg. For GM counters quenched with an organic gas, the plateau normally is 200–300 V long, with a slope not exceeding 1–2%/100 V. Halogen-quenched tubes have a shorter plateau (100–200 V) with a slope of 3–4%/100 V. The starting voltage usually is lower for halogen-quenched tubes than for tubes quenched with an organic gas.

A few GM detectors are illustrated in Figure 8-15. The end-window G–M tube (Figure 8-15, A) is used often for assay of radioactive samples. Usually, the window is split-mica thin enough to admit α particles and low-energy β particles (for example, β particles from ^{14}C and ^{35}S) into the counting volume. End-window tubes range in diameter from 1/4 to 2 in. and in window thickness from 1.4 to 5 mg/cm². Detectors with thin walls (such as 30 mg/cm²) of metal or glass are used commonly with portable survey meters (Figure 8-15, B) and for measurement of the activity of β-emitting radioactive solutions (Figure 8-15, C). The usefulness of a dipping tube for measurement of radioactive solutions is limited by self-absorption of the radiation in the volume of solution surrounding the

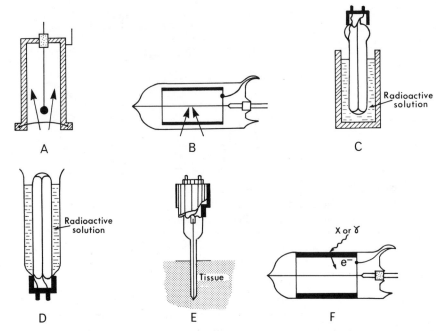

Figure 8-15 Various Geiger–Mueller detectors: (A) end-window; (B) side-window; (C) thin-wall dipping tube; (D) thin-wall jacketed tube; (E) needle probe; (F) tube with a heavy metal wall.

detector. The counter in Figure 8-15, D, may be used to measure the activity of a solution flowing through the jacket. With the needle-probe detector (Figure 8-15, E), the sensitive volume occupies the tip of a long thin probe or catheter. These probes may be either side-sensitive or tip-sensitive and may be used to measure the concentration of a β-emitting sample in tissue. For example, needle-probe detectors have been used in the diagnosis of brain, eye, and breast tumors containing increased concentrations of ^{32}P.

The efficiency of a GM counter is nearly 100% for α and β particles entering the counting volume. In a typical GM tube, x- and γ-ray photons are detected with an efficiency of only about 1%. The efficiency can be increased to 6–8% by coating the cathode with a heavy metal such as bismuth or lead (Figure 8-15, F).

Flow Counting

Low-energy β particles such as those from ^{14}C and ^{35}S are absorbed readily in the window of a conventional GM tube. Ultrathin Mylar

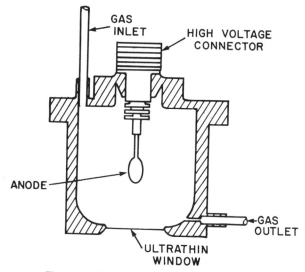

Figure 8-16 End-window flow counter.

windows of 100–200 $\mu g/cm^2$ thickness can be constructed, but they are permeable to the counting gas and must be used with a detector which receives a continuous supply (50–100 cm^3/min) of counting gas. A typical flow counter is illustrated in Figure 8-16. Methane or P-10 gas usually is employed for proportional flow counting, and a mixture of about 99% helium and 1% butane or isobutane frequently is employed for GM flow counting.

Even ultrathin Mylar windows are too thick for detection of β particles from ^3H, and a windowless flow counter is used when these low-energy radiations must be detected with a gas-filled chamber. After the tritium sample has been admitted into the sensitive volume of the detector, the air is removed by flushing with counting gas and the sample is counted. Because of the flushing operation, windowless flow counters require more gas and are slower in operation than thin-window flow counters. Also, the risks of damage to the anode and internal contamination of the detector are increased with a windowless flow counter. Some windowless flow counters use the gas emerging from the counting volume to flush the air from the environment of the next sample to be counted.

Gas-Filled Neutron Detectors

BF$_3$ Detector. Thermal neutrons may be detected with a proportional detector filled with BF$_3$ gas. The thermal neutrons undergo (n,α) reactions in the gas, and the resulting α particles and daughter nuclei produce dense ionization within the counting volume. The voltage pulses formed by neutron interactions are large and can be separated easily from pulses furnished by γ-ray interactions. Consequently, a BF_3 detector can measure thermal neutrons in the presence of γ-rays.

The cross-section is 4010 barns for the capture of thermal neutrons by the reaction

$$^{10}B(n, \alpha)^7Li$$

Although the abundance of ^{10}B is only 19.8% in natural boron, BF_3 enriched to 96% in ^{10}B is available. The count rate C_r measured with a BF_3 counter is related to the flux density ϕ of impinging thermal neutrons by the expression

$$\phi = \frac{1.128}{N\sigma} C_r$$

where N is the number of ^{10}B atoms in the counting volume and σ is the cross-section for the (n,α) reaction of thermal neutrons with ^{10}B.

Example 8-6

A thermal neutron count rate of 400 cpm was measured with a BF_3 counter 2 cm in diameter and 15 cm long. The counting gas was enriched to 96% with ^{10}B and was contained in the counter at a pressure of 100 mm Hg. The temperature was 25°C when the counts were obtained. What was the flux density of thermal neutrons?

The volume V of the chamber is

$$V = \pi r^2 h$$
$$= \pi (1 \text{ cm})^2 (15 \text{ cm})$$
$$= 47.1 \text{ cm}^3 \text{ or } 0.0471 \text{ liter}$$

The number n of moles of BF_3 in the counter is

$$n = \frac{PV}{RT}$$

$$= \left(\frac{100}{760} \text{ atm}\right)\left(\frac{0.0471 \text{ liter}}{0.082 \text{ liter-atm/mole °K}}\right) (298 \text{ K})$$

$$= 2.53 \times 10^{-4} \text{ moles}$$

The number N of ^{10}B atoms in the counter is

$$N = (0.96 \, ^{10}B \text{ atoms/molecule}) (6.02 \times 10^{23} \text{ molecules/mole})$$
$$\times (2.53 \times 10^{-4} \text{ mole})$$
$$= 1.46 \times 10^{20} \, ^{10}B \text{ atoms}$$

The flux density ϕ of thermal neutrons is

$$\phi = \frac{1.128}{N\sigma} C_r$$

$$= \frac{1.128}{(1.46 \times 10^{20} \text{ atoms})(4.010 \times 10^{-21} \text{ cm}^2/\text{atom})} \left(\frac{400 \text{ cpm}}{60 \text{ sec/min}} \right)$$

$$= 12.8 \text{ neutrons/cm}^2\text{-sec}$$

The sensitivity of a neutron detector may be defined as

$$\text{sensitivity} = \frac{\text{count rate}}{\text{flux density}}$$

The sensitivity of the BF_3 counter in this example is

$$\text{sensitivity} = \frac{400 \text{ cpm}}{(12.8 \text{ neutrons/cm}^2/\text{sec})(60 \text{ sec/min})}$$

$$= 0.52 \frac{\text{counts}}{\text{neutron-cm}^2}$$

Long Counter. A BF_3-filled proportional detector may be used also to measure fast neutrons, provided that the counting volume is surrounded with paraffin or another substance to slow (or "moderate") the fast neutrons before they reach the counting volume. To cause the detector to respond only to fast neutrons, slow neutrons may be absorbed in a thin sheet of cadmium placed outside the paraffin. If the thickness of paraffin is selected properly, the response of the detector will be independent of neutron energy from about 10 keV to more than 1 MeV. Because of the long range of energy independence, this type of detector is termed a *long counter.*

Proton Recoil Counter. Neutrons with energy greater than about 500 keV may be detected with a proportional counter which responds to protons recoiling from collisions with fast neutrons. The protons usually originate in a hydrogenous material such as methane, paraffin, or polyethylene. Thermal neutrons impinging upon the detector usually are absorbed in a thin sheet of cadmium. The sensitivity of a proton recoil counter for fast neutrons is much lower than the sensitivity of a BF_3 counter for thermal neutrons.

PROBLEMS

1. An α particle of 4.01 MeV from ^{232}Th is absorbed completely in an air-filled pulse-type ionization chamber. If the capacitance is 50 picofarad

and the induced charge reduces the effect of the charge collected by 10%, what is the size of the resulting voltage pulse?

2. A GM tube with a volume of 30 cm³ contains a counting gas of 99.9% argon and 0.1% ethyl formate at a pressure of 200 mm Hg at 25°C. If 10^8 molecules of ethyl formate are dissociated for each voltage pulse formed, what is the expected life for the tube?

3. The maximum permissible concentration in drinking water is 3×10^{-5} μCi/ml for ^{89}Sr ($\bar{E}_\beta = 0.56$ MeV) in insoluble form. What volume of water must be evaporated to measure a concentration this low in a current-type ionization chamber which can detect 10^{-13} amp, if one-half of the β particles emitted by the sample produce ionization in the counting volume?

4. For a GM detector operated at a voltage on the GM plateau:
   (a) Is the pulse produced by an α particle larger than a pulse produced by a β particle?
   (b) At a given voltage, do the size and shape of the pulse vary with the length of the anode? With the diameter of the detector?
   (c) Does the size of the pulse vary with the voltage?

5. What is the sensitivity of a BF_3 counter for thermal neutrons, if the 50 cm³ volume of the counter is filled with BF_3 (enriched to 96% with ^{10}B) at 100 mm Hg pressure at 24°C?

REFERENCES

[1] O'Kelley, G. 1962. Detection and measurement of nuclear radiation. U.S. Dep. Com., Office of Technical Services. p. 46.

[2] Hanna, G. 1959. Alpha radioactivity. *In* E. Segre (ed.), Experimental nuclear physics. John Wiley & Sons, New York. p. 192.

9

External Scintillation Detectors

Gas-filled chambers are inefficient detectors for x- and γ-radiation, because most of the x- and γ-ray photons traverse the low-density gas without interacting. To increase the probability of x- and γ-ray interactions, the gas-filled chamber may be replaced by a solid crystal with a high density and a high atomic number. Atoms of a crystal are immobile, however, and interactions cannot be registered by the collection of ion pairs.

In certain crystals (such as sodium iodide), light is released as radiation is absorbed. By directing the light onto the photosensitive surface (photocathode) of a photomultiplier tube, electrons can be ejected from the surface. The number of electrons ejected may be amplified in the photomultiplier tube to produce a measurable signal which reflects an interaction in the crystal.

Scintillation detectors may be used to detect particulate radiations as well as x- and γ-ray photons. For example, liquid scintillators are used often to detect low-energy β particles. Screens coated with ZnS are useful for the detection of α particles, and organic crystals of anthracene and stilbene are used for the measurement of α-, β-, and x-radiation. Fast neutrons are recorded occasionally by proton recoil reactions in stilbene. Organic and inorganic crystals of various types are compared in Table 9-1 and discussed in detail in the literature[1–3].

TABLE 9-1 *Properties of Organic and Inorganic Scintillators* [1]

Scintillator	Density, g/cm³	Z_{eff}	Wavelength of maximum emission, Å	Refractive index	Light yield relative to anthracene	Decay time, nsec	Remarks
Anthracene	1.25	5.8	4450	1.59	1.00	25	Large crystals not quite clear
Quaterphenyl	—	5.8	4380	—	0.85	8	Pure crystals difficult to synthesize
Stilbene	1.16	5.7	4100	1.62	0.73	7	Good crystals readily obtainable
Terphenyl (*para*)	1.12	5.8	4150	—	0.55	12	Good crystals readily obtainable
Diphenyl acetylene	1.18	5.8	3900	—	0.26–0.92	7	Large, good crystals readily obtainable
Naphthalene	1.15	5.8	3450	1.58	0.15	75	Good crystals readily obtainable
Chloroanthracene	—	9.8	—	—	0.03	—	—
ZnS(Ag)	4.1	27	4500	2.4	2.0	>1	Mostly very small crystals
CdS(Ag)	4.8	44	7600	2.5	2.0	>1	Yellow crystals
NaI(Tl)	3.67	50	4100	1.7	2.0	0.25	Excellent crystals available, hygroscopic
KI(Tl)	3.13	49	4100	1.68	0.8	>1	Excellent crystals available, not hygroscopic
NaCl(Ag)	2.17	16	2450 3850	1.54	1.15	>1	Excellent crystals available
LiI(Ti, Sn or Eu)	4.06	52	blue-green white	1.95	—	—	Activation problems, hygroscopic
CsI(Tl)	4.51	54	white	1.79	1.5	>1	Excellent crystals obtainable, not hygroscopic
CaWO₄	6.06	59	4300	1.92	1.0	>1	Small crystals, transparency good

Scintillation Detectors

Interactions of x- or γ-ray photons will not cause a crystal of pure sodium iodide to fluoresce at room temperature. However, if impurity atoms (usually thallium) are incorporated into the crystal, luminescence centers are created which can be excited by electrons released during interactions of x- or γ-ray photons. As the luminescence centers return to a more stable condition, light is released in the crystal. The importance of thallium as an "activator" for sodium iodide is illustrated in Figure 9-1.

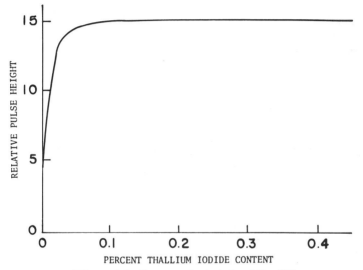

Figure 9-1 Effect of thallium on the height of the ^{137}Cs photopeak for a 1 in. diameter by 1/2 in. thick crystal of NaI(Tl). [From J. Harshaw, E. Stewart, and J. Hay. 1952. AEC Report NYO 1577.]

In most NaI(Tl) crystals, about 20–30 photons of light are released for every kiloelectron volt of energy absorbed. The light photons traverse the transparent crystal and impinge upon the photosensitive cathode (photocathode) of a photomultiplier tube. For electrons to be ejected from the photocathode, the spectral sensitivity of this component of the photomultiplier tube must correspond to the wavelength of light emitted by NaI(Tl). A photomultiplier tube with an S-11 response satisfies this criterion rather well (Figure 9-2). For every 7–10 light photons incident upon the photocathode, one electron is released. The number of electrons is multiplied at each of several stages (dynodes) in the photomultiplier tube, and a signal is furnished at the photomultiplier anode which may be

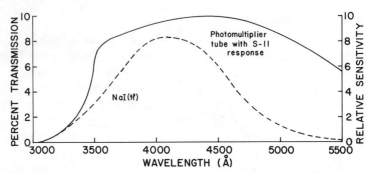

Figure 9-2 The emission spectrum of NaI(Tl) is matched closely to the spectral sensitivity of a photomultiplier tube with an S-11 response.

amplified electronically and counted. The size of the signal at the anode is proportional to the energy dissipated in the crystal by the incident radiation.

Thallium-Activated Sodium Iodide Crystals

Sodium iodide crystals suitable for γ-ray detection are prepared as single crystals with a very low concentration of potassium (and, therefore, radioactive ^{40}K). Crystals are grown in a vacuum or inert atmosphere by one of two techniques, the gradient technique and the pulling technique.

Gradient Technique. In the gradient method for growing sodium iodide crystals, molten sodium iodide in a conical vessel is lowered (for example, 1 in. per day) across a temperature gradient in a double furnace (Figure 9-3). The upper region of the furnace is maintained at a temperature above the melting point of sodium iodide (651°C), and the lower region of the furnace is maintained at a slightly lower temperature. As the tip of the conical vessel crosses the gradient between the two temperature regions, random nucleation of a single-crystal seed occurs. The crystal grows around this seed as the remainder of the vessel is lowered across the gradient. Crystals up to 16 in. in diameter may be prepared by the gradient technique. The finished crystal is released from the conical vessel by careful melting of the outer crystal layer. The crystal then is annealed at about 350°C for 2–3 hr and cooled to room temperature over a period of 20–30 hr before being machined into a cylinder of the desired size.

Pulling Technique. With this method, molten sodium iodide is maintained at a temperature several degrees above its melting point. A seed crystal attached to a rod (the "pullrod") is lowered into the melt, and

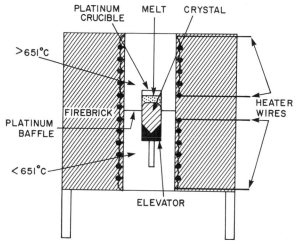

Figure 9-3 Double furnace used to prepare crystals of NaI(Tl) by the gradient technique.

crystal growth is induced by lowering the temperature of the molten sodium iodide or by increasing the removal of heat along the pullrod. After the crystal has attained the desired size, the pullrod to which the crystal is attached is withdrawn slowly from the melt. Crystals up to about 5 in. in diameter may be grown by this method.

Mounting Scintillation Detectors. Sodium iodide crystals are hygroscopic and must be protected from moisture; a crystal exposed to moisture will darken and exhibit a reduced transparency to light. Consequently, sodium iodide crystals are mounted in a dry atmosphere inside a metal canister (for example, 1/32 in. aluminum) sealed to prevent the entrance of moisture. A light pipe of Lucite, clear glass, polystyrene, or quartz may be attached to the side of the crystal nearest the photocathode of the photomultiplier tube, or the crystal may be mounted permanently on the glass envelope of the phototube. The crystal or light pipe is coupled to the photomultiplier tube with a transparent viscous medium such as silicone fluid. To increase the collection of light at the photocathode, all surfaces of the crystal except the surface adjoining the photocathode are roughened with emery paper to cause diffuse rather than specular reflection. A reflector such as aluminum oxide, magnesium oxide, or aluminum foil is packed between the outer surface of the crystal and the aluminum canister. The entire detector is made light-tight to prevent ambient light from reaching the photocathode.

Photomultiplier Tubes

A photomultiplier tube is diagrammed in Figure 9-4. The photocathode usually is an alloy of cesium and antimony; silver, magnesium, and cesium; or cesium, antimony, sodium, and potassium (bialkaki photomultiplier tube). Each of these combinations provides an acceptable number of electrons per light photon absorbed. As light impinges upon the photocathode, electrons are released and accelerated to the first dynode, a positively charged electrode positioned a short distance from the photocathode. For each electron reaching the first dynode, 3–4 electrons are released and accelerated to the second dynode, where more electrons are released. This process is repeated at 6–14 successive dynodes in the photomultiplier tube, and 10^6–10^8 electrons reach the anode for each electron liberated from the photocathode.

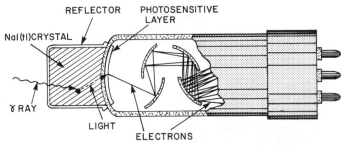

Figure 9-4   A sodium iodide crystal and photomultiplier tube.

Dynodes may be arranged in various configurations, including those referred to as "focused electrode" and "venetian blind" configurations (Figure 9-5). The amplification of the signal is very dependent upon the voltage between dynodes (usually 100–150 V), and the high-voltage supply for the photomultiplier tube must be very stable. Photomultiplier tubes should be shielded from external magnetic fields, because even weak fields influence the trajectory of electrons and reduce the amplification of the signal by as much as two-thirds. A magnetic shield of Mu-metal often is incorporated into the housing for the scintillation detector.

Electrons collected by the anode are directed into a preamplifier circuit where a pulse of a few millivolts to a few volts is formed. The size of each voltage pulse is a measure of the energy deposited in the crystal by a particular γ-ray photon.

Example 9-1

A γ-ray photon of 1.17 MeV from ^{60}Co is absorbed totally in a NaI(Tl) crystal. What size voltage pulse might be produced in a preamplifier circuit with a capacitance of 400 picofarad?

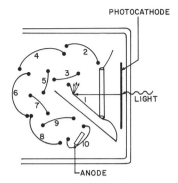
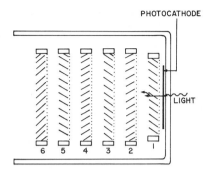

Figure 9-5 Common photomultiplier tubes. Left, focused electrode tube, and, right, venetian blind tube.

If 10 eV are required on the average for each excitation event in the crystal, then the number of excitations produced by total absorption of a 1.17 MeV photon is

$$\text{number of excitations} = \frac{1.17 \times 10^6 \text{ eV}}{10 \text{ eV/excitation}}$$

$$= 1.17 \times 10^5 \text{ excitations}$$

If 30% of the excitation events result in the release of photons of visible light, then the number of light photons released is

number of light photons

$$= (1.17 \times 10^5 \text{ excitations})(0.3 \text{ photons/excitation})$$

$$= 3.51 \times 10^4 \text{ photons}$$

If 80% of the released photons reach the photocathode, and if one electron is released from the photocathode on the average for every ten impinging light photons, then the number of electrons released is

number of electrons leaving photocathode

$$= (3.51 \times 10^4 \text{ photons})(0.8)(0.1 \text{ electron/photon})$$

$$= 2.81 \times 10^3 \text{ electrons}$$

If the amplification of the photomultiplier tube is 10^6, then the number of electrons reaching the photomultiplier anode is

$$\text{number of electrons at anode} = (2.81 \times 10^3 \text{ electrons})(10^6)$$

$$= 2.81 \times 10^9 \text{ electrons}$$

Since the charge of an electron is 1.6×10^{-19} coul, the total charge collected at the anode is

charge at anode $= (2.81 \times 10^9$ electrons$) (1.6 \times 10^{-19}$ coul/electron$)$

$$= 4.5 \times 10^{-10} \text{ coul}$$

The voltage pulse V furnished by the preamplifier circuit with a capacitance of 400 picofarad is

$$V = \frac{Q}{C}$$

$$= \frac{4.5 \times 10^{-10} \text{ coul}}{4 \times 10^{-10} \text{ farad}}$$

$$= 1.11 \text{ V}$$

A γ-ray photon of 1.33 MeV also is released by ^{60}Co. The computation above repeated for a γ-ray energy of 1.33 MeV results in an answer of 1.27 V for the size of the voltage pulse.

As the energy of impinging γ-rays increases, the detection efficiency

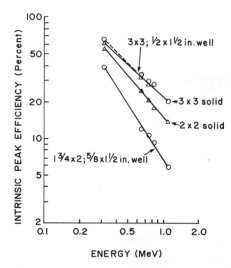

Figure 9-6 Measured intrinsic photopeak efficiency of NaI(Tl) crystals of various sizes. Sources were 9.3 cm from the face of solid crystals and at the bottom of the well for well crystals. [From C. Harris, D. Hambler, and J. Francis. Basic principles of scintillation counting for medical investigators. ORNL 2808/ORINS 30.]

of a NaI(Tl) scintillation detector decreases (Figure 9-6).* The detection efficiency may be increased by using a larger crystal and by improving the counting geometry. For example, a crystal into which the radioactive sample may be inserted (that is, a "well detector") furnishes a detection efficiency greater than that provided by a crystal of the same size which receives at best no more than half of the γ-rays from a radioactive source.

PROBLEM

1. A γ-ray photon from ^{99m}Tc (140 keV) is absorbed completely in NaI(Tl). The photomultiplier has 10 dynodes, with an average multiplication factor of 3 for electrons at each dynode. About 80% of the light released in the crystal reaches the photocathode, which has a photocathode efficiency (number of electrons emitted per light photon absorbed) of 0.05. If 20 photons of light are released per kiloelectron volt of energy absorbed, compute the number of electrons received at the anode of the photomultiplier tube.

REFERENCES

[1] Ramm, W. 1966. Scintillation detectors. *In* F. Attix and W. Roesch (ed.). Radiation dosimetry, Vol. II, 2nd ed. Academic Press, New York. p. 123.

[2] Hine, G. 1967. Sodium iodide scintillators. *In* G. Hine (ed.), Instrumentation in nuclear medicine, Vol. 1. Academic Press, New York. p. 95.

[3] Birks, J. 1964. The theory and practice of scintillation counting. MacMillan, New York.

*The detection efficiency is expressed in Figure 9-6 as the *intrinsic peak efficiency*, defined as the ratio of the voltage pulses falling within the photopeak to the number of γ-ray photons emitted by the source. Photopeaks are discussed in Chapter 12.

10

Semiconductor Radiation Detectors

Semiconductor radiation devices are being used with increasing frequency for the detection of charged particles and photons emitted by radioactive nuclei. Among the desirable properties of semiconductor detectors are:

(1) Excellent resolution
(2) Small size
(3) The formation of pulses with fast rise times
(4) A response which varies linearly with the energy deposited in the detector and does not depend upon the type of radiation depositing the energy
(5) Negligible absorption of energy in the entrance window of the detector

The mechanism of response of a semiconductor detector is similar to that for a gas-filled detector. Ionization produced within the sensitive volume of the detector is converted to a voltage pulse which is amplified and counted. The size of the voltage pulse varies with the energy expended in the detector by the incident radiation. Compared to the pulse from a gas-filled detector, the voltage pulse from a semiconductor detector is larger and is a more precise representation of the energy deposited in the detector. Also, the ionization is collected more rapidly in a semiconductor detector, and the rise time of the pulse is shorter.

134

Semiconductor Detectors

A semiconductor is a crystal with very high resistance to the flow of electric current at low temperatures. The high resistance results from the immobility of electrons due to their confinement to the valence band of the crystal. At higher temperatures, some of the electrons escape from the valence band into the conduction band. In the conduction band, electrons are free to migrate through the crystal, and the electrical resistance is reduced. For electrons to move spontaneously from the valence to the conduction band at higher temperatures, the energy difference between the valence and conduction bands must be small. For example, only 0.6 eV is required for transition of an electron from the valence to the conduction band in germanium. In an electrical insulator, an energy difference of 2–10 eV exists between the valence and conduction band, and electrons do not migrate into the conduction band even at elevated temperatures.

Semiconductors used as radiation detectors contain impurities which contribute electrical carriers to the crystal. These impurities are added by doping a pure semiconductor to produce an "impure" or "extrinsic" semiconductor. A p-type semiconductor is formed by doping a semiconductor element (for example, germanium or silicon) with an electron-acceptor impurity. For example, tetravalent germanium may be doped with trivalent boron, indium, or gallium to produce a crystal with "holes" in the crystal lattice which can be filled by nearby valence electrons. Only 0.01–0.1 eV are required for an electron in the valence band to fill a hole in a p-type semiconductor (Figure 10-1). The hole which the electron leaves in the valence band may be filled by a neighboring electron in the valence band, creating another hole which may be filled by an adjacent electron. In this manner, holes migrate in the valence band of a semiconductor just as electrons migrate in the conduction band.

An n-type semiconductor is formed by doping an element such as

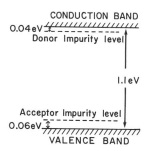

CONDUCTION BAND
0.04 eV
Donor Impurity level

1.1 eV

Acceptor Impurity level
0.06 eV
VALENCE BAND

Figure 10-1   Energy bands for electrons in silicon.

germanium with an electron-donor impurity (for example antimony) which contributes unneeded electrons to the lattice. The unneeded electrons do not participate in chemical bonding, and occupy an energy level just below the conduction band. Very little energy (0.01–0.1 eV) is required to move these electrons into the conduction band (Figure 10-1).

For a semiconductor device to be useful as a radiation detector (or, for that matter, as a diode or transistor), a junction must be formed between two dissimilar semiconductor materials (Figure 10-2). Electrons from the n-type region migrate across the junction and fill holes in the p-type region, creating an imbalance of charge, and, consequently, an electric field, across the junction. Because of the neutralization of holes by electrons, the region in the immediate vicinity of the junction is depleted of

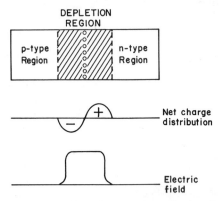

Figure 10-2 A p-n junction semiconductor detector.

mobile electrons and holes and is termed a *depletion region*. By applying a reverse bias across the junction (positive potential to the n-type region, negative potential to the p-type region), the electric field is enhanced and the depletion region is widened. If a charged particle or x- or γ-ray photon loses energy within the depletion region, electrons are raised from the valence to the conduction band, and holes are left in the valence band. These electrons and holes migrate to the appropriate positive or negative terminal of the semiconductor detector in a manner similar to the movement of electrons and positive ions in a gas-filled detector. In fact, electrons raised to the valence band, together with the holes left in the valence band, constitute the ion pairs for a semiconductor detector.

Simple germanium p-n junction detectors cannot be produced with a depletion region wider than about 1 mm. To obtain a more efficient detector, the depletion region must be widened by a process known as "drifting," in which lithium is deposited on the surface of p-type germanium by painting, vacuum evaporation, or electrodeposition. The lithium atoms are a donor impurity in germanium and silicon and, consequently, exhibit a net positive charge. Under reverse bias, the lithium atoms migrate into the p-type semiconductor and balance the electron-acceptor atoms. This balancing of the electron-acceptor impurities with the electron-donor lithium atoms creates a depletion region 12 mm or more in thickness which is confined by p-type and n-type regions on opposite sides. The lithium-drifted semiconductor acts as a simple p-n junction and, consequently, as a radiation detector. About 4 weeks are required for the lithium ions to migrate a distance of 12 mm in germanium.

Detector Cryostats

Because of the high mobility of lithium ions in germanium, Ge(Li) detectors must be operated at a temperature near $-200°C$. Detector temperatures this low are achieved by positioning the detector permanently in a cooling device or cryostat consisting of a vacuum chamber for the detector and a dewar for a cooling medium. The cooling medium almost always is liquid nitrogen.

The dewar consists of two or more concentric metal containers, with the inner container serving as the reservoir for liquid nitrogen. To reduce the transmission of heat from outer to inner container, the space between the containers is evacuated and the surfaces of the containers are highly polished.

Two cryostats, "dip stick" and "chicken feeder" cryostats, are used frequently in both vertical and horizontal configurations (Figure 10-3). With the dipstick cryostat, heat is conducted from the detector along a copper or aluminum cooling rod which is immersed in the cooling medium. The detector is oriented vertically in the vertical dipstick cryostat and horizontally in the horizontal dipstick cryostat.

In the chicken feeder cryostat, the detector is mounted at the end of a "cold finger" of liquid nitrogen maintained by gravity feed from a reservoir of liquid nitrogen above the detector. The chicken feeder cryostat has two openings and does not insulate the cooling medium from the environment as effectively as the dipstick cryostat, which has only one opening. Also, bubbles formed as the liquid nitrogen evaporates rise slowly to the surface of the medium and produce vibrations which may affect detector resolution. This difficulty can be eliminated almost completely by use of a baseline restoration circuit. Chicken feeder cryostats are used widely because they are handy and adaptable to varying experimental conditions.

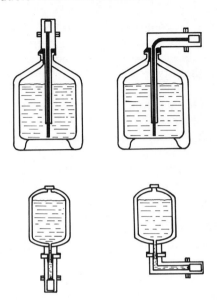

Figure 10-3   Top, dipstick cryostats in vertical (left) and horizontal (right) configuration. Bottom, chicken feeder cryostats in vertical (left) and horizontal (right) configurations. [From H. Fiedler, and O. Tench. Germanium (Li) Gamma Spectrometer Systems, Canberra Industries, Inc., Meriden, Conn.]

Response of Semiconductor Detectors

If a potential difference is applied across a pure semiconductor, a current is produced even if the semiconductor is not exposed to radiation. This current is the sum of: (1) a bulk current which is dependent upon the electrical resistance of the semiconductor; and (2) a surface current which is caused by charge leakage at the surface of the semiconductor. By cooling a Ge(Li) detector, the bulk current may be reduced to an acceptable level. The surface current can be reduced by special techniques for cleaning, etching, and encapsulating the detector.

The size V of a voltage pulse from a semiconductor detector equals the charge Q collected by the electrodes, divided by the capacitance C of the depletion region. The charge Q equals Ne, where N is the number of electron-hole pairs produced and e is the charge of the electron. The number of electron-hole pairs produced in a detector depends upon the energy deposited in the detector. Consequently, the size of a voltage pulse from a semiconductor detector is proportional to the energy deposited in the detector by an impinging particle or photon. The size of the pulse is not dependent on the LET of the radiation, because the electron-hole pairs are swept away immediately and cannot recombine. Therefore, the response of the detector reflects the energy deposited in the detector but not the type of radiation depositing the energy (Figure 10-4).

The average energy expended for each ion pair produced in a semiconductor detector is only about one-tenth (for example, 3.5 eV/IP in silicon and 2.98 eV/IP in germanium) of the energy required in a gas. Hence the number of ion pairs collected in a semiconductor detector is approximately 10 times the number collected in a gas-filled ionization chamber absorbing the same amount of energy. For example, a 2 MeV γ-ray photon absorbed completely in the depletion region of a Ge(Li) detector produces about 2×10^6 eV/(2.98 eV/IP) = 6.7×10^5 ion pairs. The same γ-ray photon produces only about 2×10^6 eV/(34 eV/IP) = 6×10^4 ion pairs when absorbed in an air-filled ionization chamber. The estimated % relative standard deviation % σ/N (Chapter 18) for the signal is

$$\frac{\%\sigma}{N} = \frac{100}{\sqrt{\text{number of ion pairs}}}$$

In the semiconductor detector, %σ/N is 0.12 for the signal produced by the 2 MeV γ-ray photon. For the gas-filled detector, %σ/N is 0.41. For the absorption of a given amount of energy, therefore, the range of pulse sizes is much narrower for a semiconductor detector than for a gas-filled

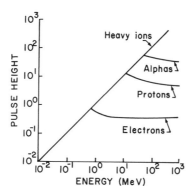

Figure 10-4 Linearity of pulse height with energy deposited by different particles in a 1 mm thick silicon detector. The detector thickness equals the particle range at 700 keV for electrons, 11 MeV for protons, and 55 MeV for α's. The pulse heights show an abrupt leveling at these energies. [From S. Friedland, and M. Zatzick. 1967. Semiconductor detectors. In G. Hine (ed.), Instrumentation in nuclear medicine. Academic Press, New York. p. 72.]

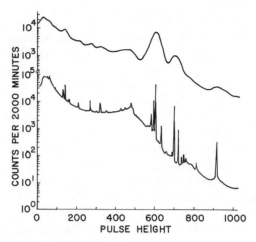

Figure 10-5 Pulse-height distributions for γ-rays from neutron-activated human lung tissue measured with a 5 in. diam × 3 in. thick NaI(Tl) crystal (upper curve) and with a 20 cm³, 11 mm thick Ge(Li) detector (lower curve). [From J. Cooper, N. Wogman, H. Palmer, and R. Perkins. 1968. The application of solid-state detectors to environmental and biological problems. Health Phys. 15:419.]

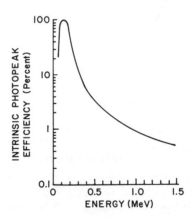

Figure 10-6 Intrinsic photopeak efficiency of a 5 cm² × 0.8 cm thick Ge(Li) detector as a function of photon energy. [From F. Goulding, and Y. Stone. 1970. Semiconductor radiation detectors. Science 170:280. Copyright 1970 by the American Association for the Advancement of Science.]

ionization chamber. Consequently, the resolution of the semiconductor detector is superior. Often, the maximum resolution obtainable with a semiconductor detector is limited by the preamplifier rather than by the detector. Pulse-height distributions obtained with NaI(Tl) and Ge(Li) detectors are compared in Figure 10-5 for γ-rays from a sample of human lung tissue activated with neutrons.

The efficiency of semiconductor detectors is nearly 100% for particulate radiations. The efficiency for detecting x- and γ-ray photons is much lower, because the depletion regions of the detectors are small. Since the atomic number is greater for germanium ($Z = 32$) than for silicon ($Z = 14$), Ge(Li) detectors exhibit a greater efficiency for detection of γ-ray photons. In Figure 10-6, the efficiency of a 5 cm^2 × 0.8 cm thick Ge(Li) detector is shown as a function of the γ-ray energy.

11

Nuclear Counting Systems

Characteristics of the signals from the radiation detectors described in Chapters 8–10 are outlined in Table 11-1. These signals may be transmitted to a variety of electronic circuits for amplification, analysis, and display. Some of the more common circuits and their applications are discussed in this chapter. By combining selected circuits and display devices, a counting system appropriate for a particular application may be developed. A general purpose counting system is outlined in Figure 11-1.

TABLE 11-1 *Characteristics of Signals from Different Detectors*

Detector	Charge collected, $coul \times 10^{-12}$	Output, mV	Pulse duration, $sec \times 10^{-6}$
Semiconductor	10^{-3}–10^{-1}	0–25	10^{-2}
NaI(Tl)	10^{-1}–10^2	0–2000	0.25
Proportional	10^{-2}–10^0	0–100	1
GM	10–10^3	0–10,000	50–300

Preamplifiers

Almost all radiation detectors have low capacitance and high impedance, and signals may be attenuated and distorted severely during transmission

142

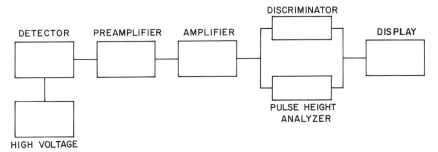

Figure 11-1 Components of a general purpose counting system.

along the high-capacitance coaxial cable between the detector and the next component of the counting system. To reduce this distortion and attenuation, a preamplifier may be located near the detector. The preamplifier is designed to match the impedance of the detector-preamplifier combination to the impedance of the cable and amplifier.

The cathode-follower preamplifier diagrammed in Figure 11-2 is one of the more common circuits for impedance matching between the detector and the counting circuit. Signals from the detector are not amplified by this circuit; in fact, the gain (gain = size of output signal/size of input signal) is slightly less than 1 for the cathode-follower preamplifier. In the cathode-follower circuit, the signal from the detector is delivered to the capacitor C which represents the total capacitance at the output of the detector, including the capacitance of the counting chamber, the capacitance of the cable between the chamber and preamplifier, and the input

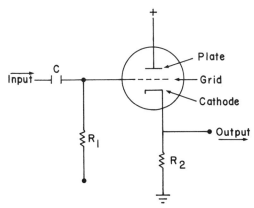

Figure 11-2 Cathode-follower preamplifier for impedance matching between the detector and the coaxial cable and amplifier.

capacitance of the preamplifier. The charge accumulated by C leaks through R_1, producing a voltage drop across R_1 which varies with the size of the input signal at C. Variations in voltage across R_1 change the bias voltage on the grid of the triode vacuum tube or on the base of a transistor used in place of the vacuum tube. The change in grid bias influences the current flowing between cathode and plate of the triode, and the voltage drop across resistance R_2 between cathode and ground varies in response to this influence. This change in voltage across R_2 constitutes the output voltage pulse for the circuit. The shape of the pulse is determined by the values of R_1 and C and by the presence of a delay line clipping circuit which may be inserted between preamplifier and amplifier to furnish a brief, symmetrical voltage pulse (Figure 11-3).

Figure 11-3   Left, original voltage pulse from a preamplifier comprised of a steep leading edge followed by a long tail. Right, brief, symmetrical voltage pulse obtained by inserting a delay line between preamplifier and amplifier.

The capacitance of a semiconductor radiation detector varies with the applied voltage and with other external conditions. These conditions, as well as the energy deposited in the detector, influence the size of voltage pulses from a semiconductor detector. The amount of charge collected by the detector varies only with the deposited energy, however, and is not influenced by the applied voltage or other external conditions. To obtain a voltage pulse which reflects only the energy deposited in the detector, a preamplifier which is charge-sensitive rather than voltage-sensitive should be used with a semiconductor detector. The charge sensitivity of a charge-sensitive preamplifier is the quotient of the output voltage from the preamplifier divided by the charge collected in the detector.

Amplifiers

Signals from a preamplifier are amplified and shaped in an amplifier. The ratio of the height of the voltage pulse leaving the amplifier to the size of the input signal is referred to as the *amplifier gain*. The gain of an amplifier may vary from 1 to 50,000, depending upon the particular detector and counting system used.

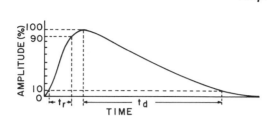

Figure 11-4 Typical pulse furnished by an amplifier. The amplitude may be either voltage or instantaneous current. The pulse rise time t_r and decay time t_d are illustrated.

A typical pulse from an amplifier is shown in Figure 11-4. The pulse rise time t_r is the time required for the pulse amplitude to increase from 10 to 90% of its maximum value. The pulse decay time is the time required for the pulse to decrease from maximum amplitude to 10% of maximum. The rise time of an amplifier should be less than the time required by the detector to form the input pulse. For a gas-filled detector, for example, the amplifier rise time should be less than the time required to collect the charge produced during interaction of radiation within the sensitive volume of the detector. To prevent the amplifier from summing successive pulses from the preamplifier, the amplifier pulse should be terminated rapidly after it reaches maximum amplitude. The integration time of an amplifier is the total time required for formation of an output pulse in the amplifier. The integration time is a compromise between the time required for amplification of the entire signal from the detector and the time which results in summation of a significant number of successive pulses. For an amplifier used with a NaI(Tl) detector, the integration time is about 1 μsec.

If a very large pulse (such as a pulse produced by interaction of a cosmic ray in the detector) is received by the amplifier, then the amplifier momentarily may become unstable. As the amplifier returns to stable operating conditions, pulses from the amplifier may be distorted. This distortion of output pulses is termed *pulse amplitude overloading*, and is particularly significant when the amplifier is operated with high gain. Pulse distortion can be caused also if pulses are received at too high a rate at the input terminal of the amplifier. This cause of distortion of output pulses is referred to as *count rate overloading*, and results in amplitude enhancement (pulse "pile up") of output pulses.

Amplifiers used in counting circuits usually are linear amplifiers. In a linear amplifier, the height of the output pulse varies linearly with the size of the input pulse. Linear amplification may be a handicap if pulses from the detector differ greatly in size. For these applications, an amplifier with logarithmic gain may be used in place of a linear amplifier. For example, logarithmic amplifiers are used frequently in liquid scintillation counters for simultaneous counting of pulses from ³H and ¹⁴C. Logarithmic amplifiers also are useful for simultaneous counting of γ-ray photons of widely different energies (for example, photons of 35 keV from ¹²⁵I and 364 keV from ¹³¹I). In a logarithmic amplifier, the size of the output pulse is proportional to the logarithm of the size of the input pulse. Consequently, a wide range of input pulses may be amplified in a logarithmic amplifier without pulse amplitude overloading or rejection of very small pulses (Figure 11-5).

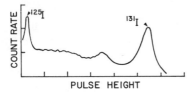 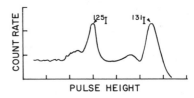

Figure 11-5 Output pulses for γ-ray photons from ¹²⁵I and ¹³¹I furnished by: left, a linear amplifier and right, a logarithmic amplifier.

Pulse-Height Analyzers

With some radiation detectors (ionization chamber, proportional, scintillation and semiconductor detectors), the size of the signal is proportional to the energy deposited in the detector by a charged particle or x- or γ-ray photon. Depicted in Figure 11-6 is a typical train of pulses from a linear amplifier connected to such a detector. These pulses may be sorted by a pulse-height analyzer to yield a pulse-height spectrum which reflects the varying amounts of energy lost in the detector by incident photons or particles. Two techniques, differential sorting and integral sorting, are employed for pulse-height analysis.

For integral sorting, a single discriminator in the pulse-height analyzer is varied from a position where all pulses are transmitted to a position where all pulses are rejected. Shown in Figure 11-7 is an integral spectrum

Figure 11-6 Train of voltage pulses provided by an amplifier connected to an ionization chamber, proportional, scintillation, or semiconductor detector. The height of each pulse reflects the energy deposited in the detector by an incident particle or photon.

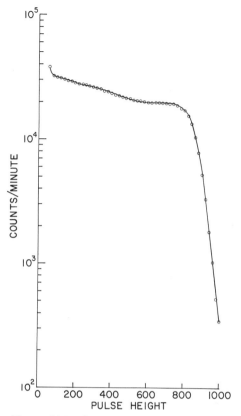

Figure 11-7 Integral spectrum for [137]Cs.

for [137]Cs. At any value of pulse height on the *x* axis, the height of the curve denotes the number of pulses which are large enough to pass the discriminator and enter the counter. The Schmitt trigger circuit is used often as the discriminator in a pulse-height analyzer[1].

A differential pulse-height analyzer is composed essentially of two discriminator circuits connected to an anticoincidence circuit (Figure 11-8). Each discriminator receives pulses simultaneously from the amplifier and transmits pulses above a predetermined minimum height. In Figure 11-9, for example, the lower discriminator transmits pulses above size V_1 and the upper discriminator transmits pulses above size V_2. Pulses of size less than V_1 are rejected by both discriminators and are not transmitted to the coincidence circuit. Pulses of size between V_1 and V_2 are transmitted by the lower discriminator and rejected by the upper discriminator; consequently, pulses are received at only one input terminal of

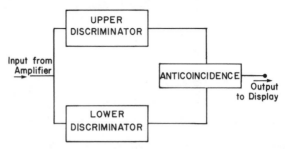

Figure 11-8 Two discriminator circuits and an anticoincidence circuit used for differential pulse-height analysis.

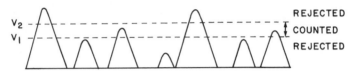

Figure 11-9 Diagram of a series of voltage pulses received by discriminators V_1 and V_2 of a differential pulse-height analyzer. Pulses with height between V_1 and V_2 are counted. Pulses with height below V_1 are rejected by both discriminators, and pulses with height greater than V_2 are rejected by the anticoincidence circuit.

the anticoincidence circuit. Pulses large enough ($> V_2$) to be transmitted by both discriminators are delivered simultaneously to both input terminals of the anticoincidence circuit. The anticoincidence circuit delivers a pulse to the display device when it receives a pulse at only one input terminal. When signals are received simultaneously at both input terminals, a pulse is not transmitted to the display device. Consequently, the display device registers only the number of pulses of size between V_1 and V_2. The range of pulse sizes recorded by the display device may be varied by changing the settings V_1 and V_2 of the lower and upper discriminators. These settings may be indicated on the pulse-height analyzer by various labels such as "lower discriminator" and "upper discriminator," "E_1" and "E_2," or "lower level" and "upper level."

Occasionally, the lower and upper discriminators of a pulse-height analyzer are not variable. With this type of analyzer, differential pulse-height analysis is achieved by changing the range of pulse sizes emerging from the detector or amplifier. With a scintillation detector, these changes may be accomplished by varying the high voltage of the photomultiplier tube or the gain of the amplifier.

In a pulse-height analyzer with discriminators which are independently adjustable, the range of pulse sizes transmitted to the anticoincidence circuit may be affected severely by small fluctuations in voltage applied to the discriminators. To reduce this variation in pulse selection, the upper discriminator may be made a "slave" to the lower discriminator. That is, the upper discriminator may be arranged to "ride" on the lower discriminator, with the effect that the relative difference in pulse sizes transmitted by the two discriminators remains unchanged. The lower discriminator may be termed "lower level," "threshold," "E" or "baseline," and the range of pulse sizes transmitted by the two discriminators may be described as "window width," "window," "slit width," or "ΔE." The position of the lower discriminator determines the minimum size of pulses to be recorded, and the width of the window determines the increment of pulse sizes to be recorded.

Most pulse-height analyzers may be operated in either the integral or

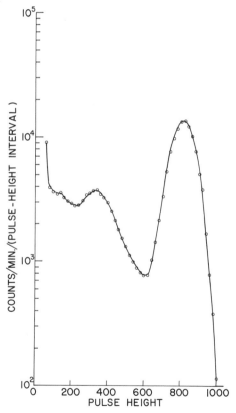

Figure 11-10 Differential spectrum for [137]Cs.

differential mode. In the differential mode, some analyzers may be operated with independent lower and upper discriminators, or with a variable lower discriminator and a dependent window. A differential pulse-height spectrum for ^{137}Cs is shown in Figure 11-10.

The linearity of a pulse-height analyzer describes the relationship between the position of the lower discriminator (or center of the window) and the size of pulses recorded by the display device. For a γ-counting system with linear amplification, a straight line should be achieved when the position of the center of the analyzer window yielding the maximum number of counts is plotted as a function of γ-ray energy (Figure 11-11).

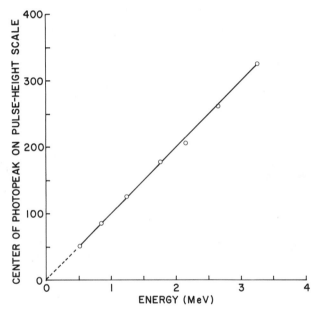

Figure 11-11   Position of the center of the analyzer window as a function of γ-ray energy. Data were obtained with a [56]Co source and NaI(Tl) scintillation detector.

The maximum departure from a straight line is termed the integral linearity and is less than 1% in a satisfactory counting system. The stability of a counting system describes the constancy in the range of pulse sizes transmitted by the analyzer window in a particular position over a long period of time and over a wide range of ambient temperatures, input count rates, and fluctuations of input voltage.

Multichannel Analyzers

With a single-channel pulse-height analyzer, the pulse-height distribution is determined by recording counts as the window of the analyzer is moved in increments across the range of pulse sizes furnished by the amplifier. This procedure is time consuming and imprecise, because each increment of pulse size is sampled independently and for only a short period. Also, pulse-height spectra for nuclides with short half-lives are difficult to measure by this technique. With a single-channel analyzer and a counting time of 1 min per channel, for example, more than 100 min are required to measure a pulse-height distribution separated into 100 parts. The same data could be collected in 1 min if a multichannel analyzer were available with 100 separate channels which collect data simultaneously. Multichannel analyzers are available commercially with approximately 100, 250, 400, 1000, 4000, and more channels.

In the multichannel analyzer diagrammed in Figure 11-12, a pulse from the amplifier is received by an analogue-digital converter (ADC). The amplitude of the pulse determines the amount of charge stored by a capacitor in the ADC. The capacitor then is discharged to a certain fraction of its stored charge while an oscillator emits pulses at a constant rate. The number of pulses emitted by the oscillator as the capacitor discharges reflects the amplitude of the original pulse received by the ADC. The number of pulses from the oscillator determines the location for the particular input pulse within the magnetic core memory where information concerning the number and amplitude of incoming pulses is stored. At each of these locations in the magnetic core memory, a binary number is

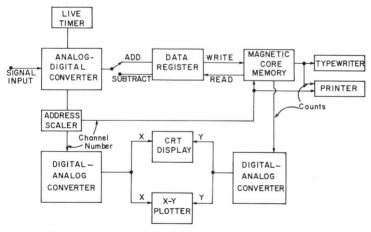

Figure 11-12   A multichannel pulse-height analyzer.

stored. At the core location appropriate for a particular input pulse, the binary number is transferred to the Add-1 Scaler, increased by 1, and then returned to its original core location. Each storage location in the magnetic core memory corresponds to a different increment in pulse size, and the binary number stored at each location reflects the number of pulses of a particular size which have been received by the ADC during the counting period.

For display of the data stored in the magnetic core memory, two digital-analogue converters (DAC) and, usually, an oscilloscope are used. One DAC relates the location being sampled in the memory to the position of the electron beam along the x axis of the oscilloscope screen. The other DAC relates the number of pulses recorded at the location being sampled to the deflection of the electron beam along the y axis of the screen. In this manner, the pulse-height spectrum is displayed almost instantaneously on the oscilloscope screen. An x-y plotter or printer also may be used for data display.

Multichannel analyzers provide various options for arithmetic manipulation of stored data. For example, a background spectrum can be subtracted automatically from a pulse-height spectrum, or the total number of counts can be determined within a particular photopeak of a spectrum. A spectrum stored in one part of memory can be transferred to another location and added or subtracted from the spectrum stored at that location. This option is useful for "spectrum stripping" operations to determine the composition of a sample containing a mixture of γ-emitting nuclides.

Scalers

A scaler is used to record or "count" the number of pulses produced in a radiation detector and transmitted by intermediate counting circuits. Early scalers were composed of a series of "flip-flop" circuits, with each circuit in either a conducting or nonconducting state at any particular moment. Each flip-flop circuit transmitted every second pulse to the next circuit. A neon bulb in each circuit indicated the state of conduction of the circuit. In this manner, the first bulb represented 2^0 (or 1) count, the second 2^1 (or 2) counts, the third 2^2 (or 4) counts, the fourth 2^3 (or 8) counts, etc. Usually, a mechanical register followed the neon bulbs and displayed counts beyond the counting capacity of the flip-flop circuits. The arrangement of neon bulbs and mechanical registers, referred to as a *binary scaler*, presented data in an inconvenient base 2 format. Most binary scalers have been replaced by scalers which display data in a base 10 format.

Data displayed by a decade or base 10 scaler may be read directly. That is, a count of 732 is displayed as the number 732 (in base 2 format this number is written 1011011100). Various methods are used to achieve a base 10 display. For example, some scalers employ a series of four binary circuits for each decade, with false counts or an inhibiting gate included to cause the display unit for the decade to register a number from 1 to 9. Other scalers utilize beam switching vacuum tubes which move an electron beam from one of ten display electrodes to the next when a pulse is received at the tube input.

Timers

Electromechanical timers driven by a synchronous motor are used with some counting systems. The timer may indicate elapsed time or may stop the accumulation of counts after a preset counting time has elapsed. Electronic timers are more accurate and are used with most newer scalers. These timers consist of a scaler pulsed by a constant-frequency oscillator and provide a visual or printed display of the elapsed or preset time.

High-Voltage Supplies

A high-voltage supply consists essentially of: (1) a transformer to convert line voltage into high voltage; (2) a rectifier to transform the ac high voltage to dc; (3) a smoothing circuit to smooth the voltage from the rectifier; and (4) a voltage stabilizer circuit. The requirements for a high-voltage supply in a counting system vary with the type of detector used. For example, a high-voltage supply for a GM tube or semiconductor detector need not be so stable as that for a proportional detector or scintillation detector, because the signal from a GM or semiconductor detector is not affected greatly by small fluctuations in applied voltage, whereas the signal from a scintillation detector may vary as much as 10% for a 1% fluctuation in high voltage. The stability of a high-voltage supply may be affected by variations in temperature, line voltage, or resistance (load) across the output terminals.

Ratemeters

Ratemeters indicate the rate at which pulses are received at the input terminal of the ratemeter. In an analog ratemeter, the receipt of a pulse at the input terminal causes the addition of a fixed quantity of charge to a tank capacitor C. The charge in the tank capacitor leaks across a resistance R at a rate which increases with the amount of charge stored in C. That is, charge is accumulated by the capacitor until the rate of charge accumulation is balanced by the rate of charge leakage. When equilibrium between charge accumulation and leakage is achieved, the voltage across the resistance *R* reflects the rate of receipt of pulses by the ratemeter. This voltage is displayed by a meter calibrated in units of count rate.

The product of resistance and capacitance of the tank circuit is referred to as the time constant *RC* for the ratemeter. The time required to attain equilibrium between charge accumulation and leakage varies with the

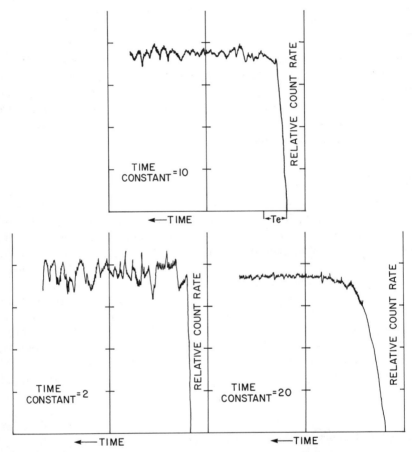

Figure 11-13 Top, graph of meter position of an analog ratemeter changing from a count rate of 0 to an average count rate *N*. The equilibrium time T_e for the ratemeter is shown. Bottom (left), rapid response and high sensitivity to statistical fluctuations in count rate are provided by a short time constant. Bottom (right), data-smoothing but a slow response are furnished by a long time constant.

input pulse rate and the time constant. The equilibrium time is the time required for an analogue ratemeter to increase from 0 count rate to within one standard deviation of the equilibrium count rate. The equilibrium time T_e is

$$T_e = RC\{0.5 \ln [2N (RC)]\} + 0.394$$

where RC is the time constant and N is the equilibrium count rate [2]. The equilibrium time indicates the speed of response of an analog ratemeter (Figure 11-13).

The equilibrium time may be decreased by reducing the time constant of a ratemeter. However, this procedure increases the sensitivity of the ratemeter to fluctuations in count rate. Most ratemeters offer a selection of time constants. The time constant chosen for a particular application represents a compromise between data-smoothing obtained with a long time constant and rapid response to changing count rates provided by a short time constant.

A digital ratemeter usually consists of a timer and a buffered scaler. The scaler accumulates counts over a preset interval of time, then transfers the accumulated counts to the buffer. Almost instantly (within 50 μsec), the scaler resets to 0 and begins to accumulate counts again. The counts stored in the buffer may be displayed visually, printed or punched on paper tape, recorded on magnetic tape, or used to modulate the brightness of the light image from a cathode-ray tube.

PROBLEM

1. Compute the equilibrium time of an analog ratemeter with a time constant of 5 sec, for an equilibrium count rate of 2400 cpm.

REFERENCES

[1] Schmitt, O. 1938. A thermionic trigger. J. Sci. Instr. 15:24.
[2] Low, F. 1967. Basic considerations in nuclear instrumentation. *In* G. Hine (ed.), Instrumentation in nuclear medicine Vol. 1. Academic Press, New York. p. 29.

12

Scintillation Spectrometry

Linear attenuation coefficients of photoelectric, Compton and pair-production interactions of γ-ray photons in NaI(Tl) are plotted in Figure 12-1 as a function of γ-ray energy. For γ-ray photons of energy less than approximately 200 keV, photoelectric absorption is the most probable interaction in NaI(Tl). Compton scattering is the dominant interaction between about 200 keV and 7 MeV, and pair production is the most probable interaction above 7 MeV.

Linear attenuation coefficients for interaction of γ-ray photons in germanium are plotted in Figure 12-2. For γ-ray photons of energy less than approximately 150 keV, photoelectric absorption is the most probable interaction in germanium. Compton scattering is the dominant interaction between about 150 keV and 8 MeV, and pair production is the most probable interaction above 8 MeV.

Part or all of the energy of an x- or γ-ray photon may be deposited in a NaI(Tl) or semiconductor detector. The size of each pulse from the detector is proportional to the energy deposited in the detector by a particular γ-ray photon. Pulses from the detector may be sorted electronically to yield a pulse-height spectrum which reflects the distribution in energy deposited by γ-rays incident upon the detector. For a particular γ-emitting source, the pulse-height spectrum consists of peaks and valleys which reflect the energy of the γ-ray photons from the source, the interactions which the photons undergo with the detector and with materials surrounding the detector, and the operating characteristics of components of the counting system. If the amplifier gain or photomultiplier

156

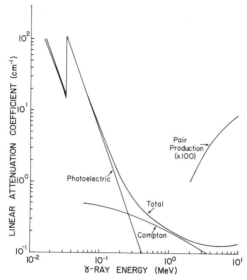

Figure 12-1 Linear attenuation coefficients of NaI(Tl) for photoelectric absorption, Compton scattering, pair production, and total attenuation, plotted as a function of γ-ray energy in MeV.

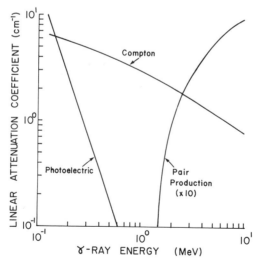

Figure 12-2 Linear attenuation coefficients of germanium for photoelectric absorption, Compton scattering, and pair production, plotted as a function of γ-ray energy in MeV[1].

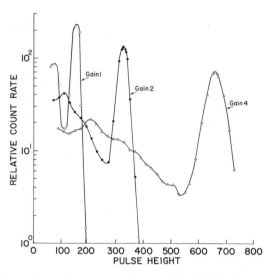

Figure 12-3 Pulse-height spectra for ^{137}Cs (662 keV γ-ray), measured with a 2 × 2 in. NaI(Tl) detector and different settings of amplifier gain.

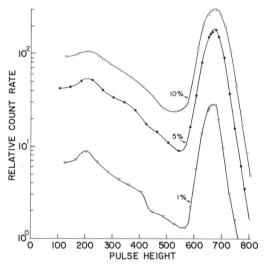

Figure 12-4 Increased count and reduced resolution are obtained as the window of a pulse-height analyzer is widened. Window widths are expressed as a percent of the highest setting of the pulse-height scale. Spectra were measured for a ^{137}Cs source in a 2 × 2 in. NaI(Tl) well detector.

high voltage is increased, then the amplification of each pulse is increased proportionately, and pulses are displayed at positions farther along the pulse-height scale. Hence, an increase in high voltage or gain expands the pulse-height spectrum over a greater range of pulse heights (Figure 12-3). At higher settings of high voltage or gain, the height of the pulse-height spectrum decreases because the voltage pulses encompass a wider range of pulse heights, and each pulse-height interval contains fewer pulses. The influence of amplifier gain and photomultiplier high voltage on the position of the photopeak along the pulse-height scale is discussed below.

As the window of the pulse-height analyzer is widened, more pulses are transmitted to the display device for each position of the window along the pulse-height scale. Consequently, the count rate for each window position increases with the width of the window. With a wider window, however, peaks and valleys in the spectrum are defined less precisely, and the resolution of the spectrum is reduced. Pulse-height spectra for ^{137}Cs measured with analyzer windows of different widths are shown in Figure 12-4.

Photopeaks

In a pulse-height spectrum, a photopeak is comprised of pulses resulting from total deposition of the energy of a γ-ray photon in the detector. The energy may have been deposited during one interaction (for example, a photoelectric interaction with subsequent absorption of characteristic x-rays) or during several interactions (such as several Compton interactions followed by photoelectric absorption of the degraded photon). The position of a photopeak along the pulse-height scale varies with the energy of the γ-ray photon and with the amplifier gain and photomultiplier high voltage. The influence of gain and high voltage on the position of the photopeak is illustrated in Figure 12-5.

The photofraction or peak-to-total ratio for a pulse-height spectrum is the quotient of the area enclosed within the photopeak divided by the total area enclosed by the pulse-height spectrum (Figure 12-6). The photofraction decreases with increasing γ-ray energy, because photons of higher energy are more likely to escape from the detector after only partial deposition of their energy (Figure 12-7).

The intrinsic peak efficiency of a detector is the ratio of the number of pulses enclosed within a photopeak divided by the number of γ-rays of corresponding energy impinging upon the detector.

The lateral spread of a photopeak is primarily a result of statistical fluctuations in processes occurring in the detector and associated electronics. For a NaI(Tl) scintillation detector, these processes include:

(a) Conversion of the energy deposited in the crystal to light
(b) Collection of light photons by the photocathode
(c) Conversion of the energy of light photons to kinetic energy of photocathode electrons
(d) Focusing of photocathode electrons onto the first dynode
(e) Electron multiplication of each dynode

Photopeak resolution may also be affected by characteristics of the electronic components of the counting system. The resolution of a pulse-height spectrum sometimes is described as the *full width at half-maximum* (FWHM), where

$$\text{FWHM} = \frac{\text{width of photopeak at half its maximum height}}{\text{position of photopeak along pulse-height scale}}$$

The FWHM usually is reported for the 0.662 MeV γ-ray photon from ^{137}Cs. For a 3×3 in. NaI(Tl) detector, the FWHM might vary from 7 to 9% (Figure 12-8).

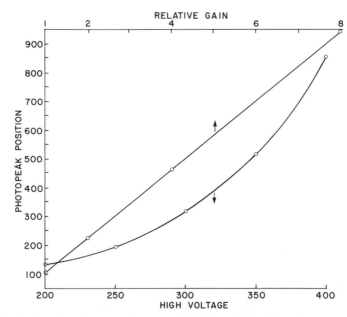

Figure 12-5 Influence of photomultiplier high voltage (lower curve) and amplifier gain (upper curve) on the position of a photopeak along the pulse-height scale. Data were measured with a ^{137}Cs source and a 2×2 in. NaI(Tl) well detector.

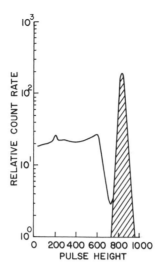

Figure 12-6 Illustration of the photofraction in a pulse-height
spectrum for ^{54}Mn. Data were measured with a 3 × 3 in. NaI(Tl)
detector. The photofraction is the quotient of the shaded area
divided by the total area under the pulse-height spectrum.
[Figures 12-6 and 12-7, from R. Heath. 1964. Scintillation
spectrometry, Vol. 1, 2nd ed. Clearinghouse for Federal Scientific
and Technical Information. Springfield, Va.]

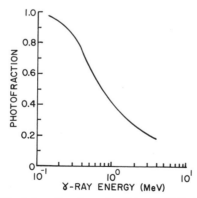

Figure 12-7 Photofraction for a 3 × 3 in. NaI(Tl) detector as a
function of γ-ray energy.

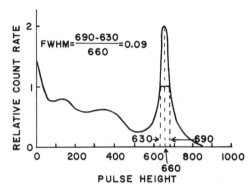

Figure 12-8 Photopeak resolution described as the FWHM.

Compton Valley, Edge, and Plateau

Compton interactions of γ-ray photons in a NaI(Tl) or semiconductor detector, with escape of the scattered photons from the detector, result in the formation of voltage pulses which are smaller than those enclosed

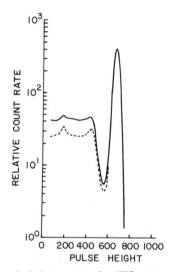

Figure 12-9 Pulse-height spectra for ^{137}Cs measured with 3×3 in. (solid curve) and 5×5 in. (dashed curve) NaI(Tl) crystals. The photopeaks are normalized to the same height and area.
[Figures 12-9 through 12-15 from R. Heath. 1964. Scintillation spectrometry, Vol. 1, 2nd ed. Clearinghouse for Federal Scientific and Technical Information, Springfield, Va.]

within the photopeak. The region of a pulse-height spectrum below a photopeak is referred to as the *Compton region* or *Compton plateau*. For a γ-ray photon of a particular energy, the relative heights of the photopeak and the Compton plateau vary with the size of the detector. In a large detector, more photons are absorbed totally in the detector. Hence, the height of the photopeak is increased and the height of the Compton plateau is reduced (Figure 12-9).

X-Ray Escape Peaks

Photons below about 200 keV are absorbed in NaI(Tl) almost entirely by photoelectric interaction. Usually, the interaction occurs with an electron in the K shell of iodine. As the vacancy created by the photoelectron is filled, one or more characteristic x-rays are released with a total energy equal to the binding energy of the K electron. If the characteristic x-rays interact in the crystal, then the light released contributes to formation of a voltage pulse which falls within the photopeak. However, if the photoelectric interaction occurs near the surface of the crystal, then the x-rays

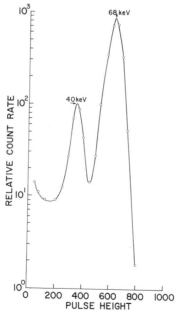

Figure 12-10 Pulse-height spectrum for ^{61}Co, illustrating the x-ray escape peak (40 keV) at a position 28 keV below the 68 keV photopeak.

may escape from the crystal without interacting. In these cases, the voltage pulse for the photoelectric interaction will be smaller by an amount equivalent to the energy carried out of the detector by the escaping x-rays. Usually, the energy loss corresponds to the escape of the 28 keV x-ray released as an electron moves from the L to the K shell in iodine. Consequently, a peak will appear in the pulse-height spectrum at a position 28 keV below the photopeak. This peak is termed an x-ray *escape peak* (Figure 12-10). With NaI(Tl) detectors, x-ray escape peaks are not observed for γ-rays with energy above about 200 keV, because the lateral spread of the photopeak for photons of higher energy obscures the presence of the escape peak. Also, higher energy γ-rays tend to penetrate farther into the detector before interaction, and the characteristic x-rays have a lower probability of escape. The reduced probability of photo-electric interaction with increased γ-ray energy also contributes to suppression of x-ray escape peaks. For γ-ray photons of energy less than 33.2 keV, the binding energy of K electrons in iodine, x-ray escape peaks are not observed, because these photons cannot interact photo-electrically with the iodine K electrons.

Characteristic X-Ray Peak

As γ-ray photons from a radioactive source interact with nearby materials such as the lead shield around the source and detector, characteristic x-rays may be released. These x-rays may be absorbed by the detector, producing characteristic x-ray peaks in the pulse-height spectrum. In Figure 12-11, the peak at 0.072 MeV reflects the absorption of characteristic x-rays from a lead shield surrounding a ^{51}Cr source and scintillation detector. The number of characteristic x-rays reaching the detector may be reduced by increasing the distance between the detector and shield, or by lining the shield with materials with an atomic number less than 82, the atomic number of lead.

Backscatter Peak

If a photon with energy greater than 200 keV is scattered at a wide angle (> 120°) during a Compton interaction, then the energy of the scattered photon is about 200 keV irrespective of the energy of the primary γ-ray. During Compton interactions of primary γ-rays in the shield around source and detector, photons may be scattered at wide angles and impinge upon the detector. These photons produce a peak at about 200 keV in the pulse-height spectrum. This peak, termed a *backscatter peak*, may be suppressed by increasing the distance between the shield and the

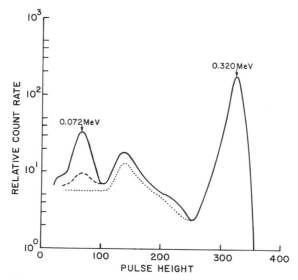

Figure 12-11   Characteristic x-ray peak at 0.072 MeV produced by characteristic x-rays from a 6 × 6 in. lead shield surrounding a NaI(Tl) detector and ^{51}Cr source. The x-ray peak may be reduced by increasing the inner dimensions of the shield to 32 × 32 in. (dotted curve) or by lining the shield with a material of atomic number less than that for lead. For example, the dashed curve was obtained by lining the 6 × 6 in. lead shield with 0.03 in. of cadmium. The 32 × 32 in. shield was lined with 0.03 in. of cadmium and 0.015 in. of copper.

detector or by choosing a material for the shield in which the number of Compton interactions is reduced (Figure 12-12).

Annihilation Peak

Two annihilation photons of 511 keV each are released during annihilation of an electron-positron pair following interaction of a high-energy γ-ray by pair production. If the pair-production interaction occurs in the detector-source shield, then one of the annihilation photons may escape from the shield and interact in the detector. The absorption of the annihilation photon furnishes a peak at 511 keV in the pulse-height spectrum. In Figure 12-13, the annihilation peak at 511 keV reflects pair-production absorption in the lead shield of 2.16 MeV γ-ray photons from ^{38}K. An annihilation peak is present also in the pulse-height spectrum for a nuclide which decays by positron emission, because two 511 keV photons are released when a positron is annihilated.

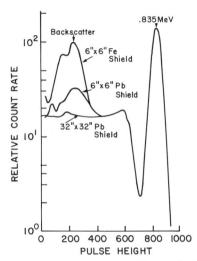

Figure 12-12 Photons scattered at a wide angle during Compton interactions of primary γ-rays in the detector-source shield may be absorbed in the detector. These photons have an energy of about 200 keV and contribute to the backscatter peak. The height of the peak is greatest for the 6 × 6 in. iron shield, because the probability of Compton interaction of 0.835 MeV γ-rays from ^{54}Mn is greater in iron than in lead The backscatter peak also is smaller for the large (32 × 32 in.) shield, because fewer scattered photons are intercepted by the detector.

Annihilation Escape Peaks

Pair-production interaction of a high-energy (> 1.02 MeV) photon in a detector is accompanied by production of two 511 keV annihilation photons, one or both of which may escape from the crystal. Voltage pulses which reflect the loss of one annihilation photon contribute to the single-escape peak which occurs at a location 0.51 MeV below the photo-peak. Pulses which reflect the loss of both annihilation photons contribute to the double-escape peak which occurs at a location 1.02 MeV below the photopeak. Single- and double-escape peaks are included in the pulse-height spectrum for ^{38}K in Figure 12-13.

Sum Peak

Many radioactive sources decay with essentially simultaneous emission of two or more photons. For example, two γ-ray photons may be emitted in cascade, or the emission of a γ-ray photon may be accompanied by

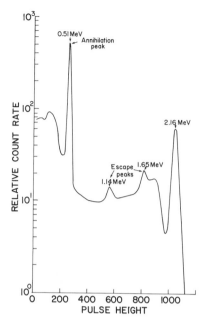

Figure 12-13 Pulse-height spectrum for ^{38}K, illustrating single- and double-escape peaks and an annihilation peak at 0.511 MeV.

release of an annihilation photon or a characteristic x-ray. If the two photons are absorbed in the detector, then one voltage pulse is produced which reflects the total energy deposited in the crystal by both photons. This pulse contributes to a sum peak or coincidence peak in the pulse-height spectrum. For example, the sum peak in Figure 12-14 reflects the simultaneous absorption of γ-ray photons of 1.17 and 1.33 MeV emitted in cascade from a ^{60}Co source. By increasing the distance between the source and the detector, the relative height of a sum peak may be reduced, because this procedure reduces the probability that both photons reach the detector.

Bremsstrahlung

Bremsstrahlung is released as β particles from a radioactive source are absorbed in the source, the source-detector shield, or the material covering the detector. If the ratio of γ-rays to β particles emitted by the source is low, and if the β particles are reasonably energetic, then the contribution of bremsstrahlung to the pulse-height spectrum is noticeable, particularly at the lower end of the spectrum (Figure 12-15).

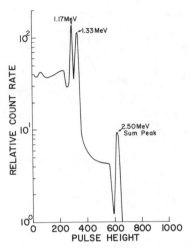

Figure 12-14 Sum peak representing simultaneous absorption in the detector of γ-ray photons of 1.17 MeV and 1.33 MeV from ^{60}Co.

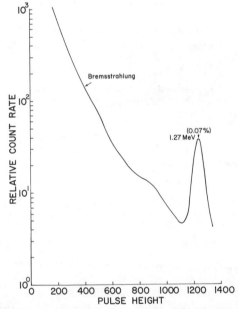

Figure 12-15 Pulse-height spectrum for ^{31}Si, illustrating the contribution of bremsstrahlung to the pulse-height spectrum for a nuclide in which the ratio of γ-rays to β particles is low (0.07%) and in which the β particles are reasonably energetic ($E_{max} = 1.48$ MeV).

Photopeak Counting

Radioactive sources which emit γ-ray photons often are counted by transmitting to the display device only those pulses which are enclosed within the photopeak. Other pulses are rejected because they are larger or smaller than those accepted by the window of the pulse height analyzer. To center the photopeak in the window, the window is positioned at a desired location along the pulse-height scale, and the amplifier gain or high voltage is varied until a maximum count rate is obtained. For example, if the center of the photopeak for ^{137}Cs is desired at 662 on the pulse-height scale, then the window is centered at 662 and the gain or high voltage is varied until the count rate is maximum. If the counting system is linear (that is, if the size of the voltage pulses varies linearly with energy deposited in the detector), then this procedure furnishes a pulse-height scale calibrated from 0 to 1000 keV, with each of the 1000 divisions on the pulse-height

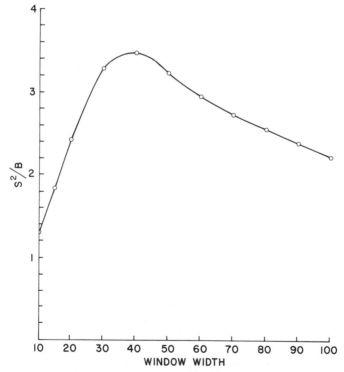

Figure 12-16 Plot of S^2/B for a ^{99m}Tc source counted with a 2 × 2 in. NaI(Tl) well detector and a single-channel pulse-height analyzer.

scale representing 1 keV. The pulse-height scale may be calibrated from 0 to 2000 keV by centering the ^{137}Cs photopeak at 331, and from 0 to 500 keV by centering the photopeak for a γ-ray photon of lower energy at a location on the pulse-height scale corresponding to twice the γ-ray energy. For example, a pulse-height scale may be calibrated from 0 to 500 keV by centering the photopeak for ^{51}Cr (γ-ray energy of 322 keV) at 644 on the pulse-height scale.

The width of the analyzer window chosen for photopeak counting depends upon the relative heights of the photopeak and the background spectrum at the location of the window. For many counting situations, the background spectrum has relatively little structure and is much lower than the photopeak, and a window should be used which just encompasses the photopeak. If the photopeak is not much greater than background, then the net count rate S for the sample should be determined for different widths of the analyzer window. Then the ratio $S^2:B$ — that is (net sample count rate)^2:(background count rate) — should be plotted as a function of

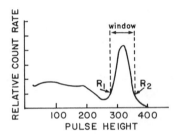

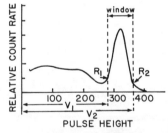

Figure 12-17 Top, for protection against a drifting window, the window of the pulse-height analyzer should be adjusted so that the lower and upper discriminators intersect the pulse-height spectrum at equal count rates ($R_1 = R_2$) near the base of the photopeak. Bottom, for protection against unstable amplification, the lower and upper discriminators should be positioned so that $R_1/R_2 = V_2/V_1$.

window width (Figure 12-16). At the greatest value for $S^2:B$, the fractional standard deviation of the net count rate for the sample is reduced to a minimum. That is, the optimum window for a photopeak not much greater than background is the window which furnishes the greatest value for $S^2:B$.

Ross and Harris[2] have suggested that if the position of the analyzer window is unstable, then the window should be widened to permit the lower and upper discriminators to intersect the pulse-height spectrum at equal heights near the base of the photopeak (Figure 12-17). If the gain of the amplifier is unstable, then the photopeak should be intercepted by the lower and upper discriminators at relative count rates R_1 and R_2, respectively, where R_1 and R_2 are determined from the expression $R_1/R_2 = V_2/V_1$, and V_1 and V_2 are the respective positions of the lower and upper discriminators along the pulse-height scale.

PROBLEMS

1. The position of the photopeak for ^{137}Cs γ-rays (0.66 MeV) changes with high voltage as shown below. The amplifier gain is 4.

position of photopeak on pulse-height scale	560	620	690	770	860
high voltage across photomultiplier tube	660	680	700	720	740

If the gain is reduced to 2, what high voltage should be selected to position the 1.38 MeV photopeak for ^{24}Na at 690?

2. A pulse-height spectrum for ^{198}Au (γ-ray of 0.412 MeV) was obtained with a scintillation spectrometer calibrated from 0 to 500 keV. A 2% window was used. Describe changes in the spectrum if:

   (a) A 5% window were used
   (b) The amplifier gain were twice as great
   (c) The amplifier gain were half as great
   (d) The upper discriminator of the window were removed

3. Describe changes in the pulse-height spectrum for ^{7}Be (γ-ray of 0.478 MeV) if:

   (a) A larger NaI(Tl) detector were used

(b) A 1 in. slab of Lucite were interposed between the ^{7}Be source and the detector

(c) A 1 in. slab of Lucite were placed behind the ^{7}Be source

4. ^{37}S decays to ^{37}Cl by β decay with a $T_{1/2}$ of 5 min. The pathway for 90% of the decay processes includes an isomeric transition of 3.13 MeV, usually achieved by emission of a 3.13 MeV γ-ray photon. A pulse-height spectrum for ^{37}S contains peaks at 3.13 MeV, 2.62 MeV, 2.11 MeV, 0.51 MeV, and about 0.20 MeV. Explain the origin of each of these peaks.

5. ^{24}Na often decays by emission of 2.76 and 1.38 MeV γ-ray photons in cascade. On the pulse-height spectrum for this nuclide, a small peak is present at 4.14 MeV. Explain the origin of this peak.

6. On a pulse-height spectrum for ^{47}Sc (γ-ray energy of 0.155 MeV), a small peak is present at 0.127 MeV. Explain the origin of this peak.

REFERENCES

[1] French, W., R. LaShure, and J. Curran. 1969. Lithium-drifted germanium detectors. Am. J. Phys. 37:11.

[2] Ross, D., and C. Harris. 1968. Measurement of clinical radioactivity, ORNL-4153. Office of Technical Information, USAEC, Washington, D.C.

13

Sample Preparation for External Counting

The choice of a technique for preparing a radioactive sample is influenced by the type and energy of the radiation emitted by the sample and by the type of detector used to count the sample. For example, low-energy β particles may be absorbed severely in a sample containing ^{14}C or ^{35}S, and the technique for preparing ^{14}C and ^{35}S samples should be chosen with this problem in mind. Samples which emit energetic γ-rays can be prepared with less concern for self-absorption. The preparation of samples for liquid scintillation counting is a complex topic which is discussed separately in the next chapter.

Dilution

Occasionally, a sample may be prepared for counting simply by dilution to a desired volume. Biologic samples which sometimes are prepared by dilution include radioactive blood, urine, milk, and saliva. If the samples emit low-energy β particles, then self-absorption may interfere with the use of the dilution method.

Ashing

Ashing techniques are employed frequently to separate a radioactive nuclide from a large volume of organic material. For example, biologic tissues containing a radioactive isotope often are prepared for counting by dry ashing or wet ashing techniques. In many cases, a sample may be dry ashed by heating to about 600°C in a muffle furnace. To prevent spattering of the sample, the furnace should be at room temperature when the sample is inserted, and then the temperature should be increased gradually. Spattering may also be reduced by drying the sample before ashing. If the sample volume is small, the sample may be ashed directly in the vessel or planchet used for counting. Otherwise, the sample should be ashed in a tared crucible and then transferred to a planchet for counting.

Dry ashing sometimes causes volatilization of the sample or fusion of the sample with the container, resulting in a loss of detectable radioactivity or in difficulties during transfer of the sample to the container used for counting. These problems are not encountered when the sample is prepared by wet ashing. Samples usually are wet ashed by adding a concentrated acid to the sample in combination with oxidizing agents and catalysts. A procedure used frequently is the Kjeldahl method, which employs concentrated sulfuric acid in conjunction with potassium sulfate, phosphoric acid, and copper or mercury salts. Compared to dry ashing, wet ashing usually requires less time for preparation of counting samples.

Digestion

Certain biologic samples can be converted to a homogeneous solution by digestion in a suitable reagent. For example, samples with limited amounts of fat can be dissolved in concentrated nitric acid. Samples with larger concentrations of fat can be separated into aqueous and fatty phases by addition of isoamyl alcohol to the mixture of sample and nitric acid. Acetone or dioxane added to this mixture furnishes a common solvent for the two phases, with the solvent comprising 60–80% of the final solution. Animal tissues other than bone may also be digested by soaking for 1–2 hr in hot formamide.

Combustion

By employment of an appropriate combustion technique, most biologic samples containing ^{14}C or ^{3}H can be oxidized completely. The ^{3}H is converted to ^{3}H-water and the ^{14}C is released as $^{14}CO_2$ which may be assayed as a gas or converted to a solid such as $Ba^{14}CO_3$. For wet combustion of a sample, a device similar to the modified Van Slyke–Folch apparatus diagrammed in Figure 13-1 may be used. After NaOH has been placed in one flask and the sample in the other flask, the apparatus is evacuated and sealed. Then the combustion solution (fuming sulfuric acid, phosphoric acid, and chromic trioxide) is introduced into the sample flask and the mixture is heated slowly for about 10 min. After the $^{14}CO_2$ released during combustion has attained equilibrium with the NaOH absorbent (about 1/2 hour), a 1 M solution of $BaCl_2$–NH_4Cl is introduced into the absorbent flask. The $BaCO_3$ precipitant is collected by filtration or centrifugation and transferred to planchets for counting.

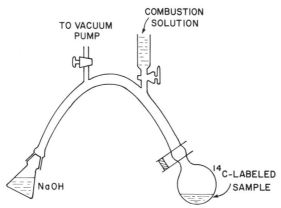

Figure 13-1 Modified Van Slyke–Folch apparatus for wet combustion of samples containing ^{14}C. [From C. Wang, and D. Willis. 1965. Radiotracer methodology in biological science. Prentice-Hall, Englewood Cliffs, N.J.]

Other methods of wet oxidation include the use of sulfuric acid and potassium persulfate with silver nitrate catalyst for ^{14}C-labeled samples soluble in water[1], and the use of a mixture of nitric and perchloric acid for the oxidation of ^3H-labeled samples to ^3H-water[2].

Dry combustion of ^{14}C- or ^3H-labeled material in an oxygen atmosphere often is used to prepare samples for assay by an ionization chamber or liquid scintillation technique[3–6]. In a modification of the Schöniger oxygen flask method of combustion, a sample containing up to 300 mg of combustible material is placed in a plastic or cellophane bag and dried. The bag is suspended in a platinum or nichrome basket and the assembly is placed in a leak-proof flask (such as a 2 liter Erlenmeyer flask) containing oxygen. The sample is oxidized by electrical current supplied to the basket by an ignition unit, or by heat furnished by an external infrared lamp. Tritiated water released during combustion may be captured as ice in a trap immersed in dry ice and acetone or cellosolve, and $^{14}CO_2$ may be retained in an alkaline absorbing solution introduced through a side arm in the reaction vessel. To reduce the possibility of explosion, the sample must be heated gradually in an atmosphere absolutely free of organic vapors[7, 8].

The Parr oxygen bomb has been used to combust relatively large samples (up to a few grams) of labeled material. Activities as low as $4 \times 10^{-4} \mu Ci$ of ^3H and $1 \times 10^{-4} \mu Ci$ of ^{14}C in 3 g of tissue have been reported[9].

For small samples (< 25 mg), dry combustion may be achieved by sealing the sample and an inorganic oxidizing agent (for example, CuO or NiO) in a Pyrex or Vycor tube. The tube is heated and then broken, and the oxidation products are transferred by vacuum line to an absorbent or to a gas-flow detector[10–12]. An alternate method involves oxidation of the sample with an inorganic oxide in a combustion chamber furnished with a steady flow of oxygen[13–15].

Evaporation of Volatile Solvents

A sample suspended or dissolved in a volatile liquid (such as water, alcohol, ether, or acetic acid) may be converted to a solid sample by evaporating the liquid. In this procedure, an aliquot of the radioactive solution is pipetted onto a counting planchet, and the liquid is evaporated gently in an explosion-proof oven or under a heat lamp. The sample should be heated carefully to prevent evaporation of the sample. The loss of ^{14}C in labeled carbonates by exchange with atmospheric CO_2 may be reduced by evaporating the liquid in a CO_2-free stream of air.

The usefulness of the solvent evaporation technique for sample preparation often is limited by the uneven thickness of the radioactive residue. Thickness uniformity sometimes may be improved by pipetting the liquid sample onto a piece of lens paper placed in the planchet. To prevent the paper from curling during evaporation, a drop of a solution which does not mix with the sample and which acts as a glue when dried may be deposited between the paper and the planchet. For nonaqueous samples, aqueous sucrose solution often is used. If the upper layers of paper draw solvent from lower layers, then the sample may concentrate near the top of the paper and self-absorption and backscatter corrections may be difficult to estimate. This source of error may be identified by counting the sample twice, with the paper inverted in the planchet between counts. Sample uniformity may be improved by rotating the sample during evaporation, by using a wetting agent (such as tetraethylene glycol) on the planchet surface, or by applying successive drops of radioactive solution to the planchet between evaporations. Gelling agents such as agar and carboxymethylcellulose have been used to immobilize the sample at a constant thickness during evaporation.

Preparation of Liquid Samples

Procedures for obtaining samples of uniform thickness during evaporation of a volatile solvent also are useful for preparing liquid samples of relatively small volume. For liquid samples of greater volume, the liquid may concentrate along the periphery of a cupped planchet or creep over the edge of a flat counting disk. To restrict the sample to the center of the planchet or disk, the periphery may be coated with silicone grease, oil dissolved in chloroform or ether, or Zapon lacquer. These materials provide a barrier to migration of most samples. Planchets are available commercially with bottoms comprised of concentric raised rings. A liquid sample added to the center of a ringed planchet fills the center depression first, then overflows into the second, third, and remaining depressions. A 1 ml sample fills all depressions to equal thickness. The use of a ringed planchet prevents the concentration of a liquid sample along the periphery of the sample container.

Preparation of Dry Powder Samples

A radioactive sample in the form of dry powder can be spread to fairly uniform thickness in a cupped planchet and then compacted by tapping the planchet with a mechanical or electric vibrator. This method is more suitable for relatively thick samples, because uniform thickness is difficult to achieve with thin samples. Powder samples can be stabilized by adding a few drops of a binding agent such as collodion (0.5% in acetone) or isobutyl methacrylate polymer (2% in ethyl alcohol). Samples of dry radioactive powder can be compressed into a planchet with a compressing plunger or into a wafer with a pilling press. The stability of a powdered sample sometimes may be improved by sintering the sample in an oven.

Filtration of Precipitates

Counting samples often are prepared by converting a radioactive solution into a suspension of radioactive particles by precipitation (for example, ^{14}C as barium carbonate, ^{35}S as barium or benzidine sulfate, and ^{45}Ca as calcium oxalate). This approach is particularly useful for solutions of small amounts of radioactivity in large volumes of solvent. After addition of a precipitating agent, the suspension usually is filtered in a Büchner or Gooch filter in which the filter disk serves as the source mount. The filter disk may be either flexible (glass-fiber filter paper or Millipore filter disk) or rigid (porous glass or sintered metal disk). To improve the uniformity of pore size, the filter disk may require a coating of fine asbestos, fuller's earth or dratomaceous earth. The coating is applied by inserting the filter in the filter assembly (Figure 13-2), pipetting or pouring into the assembly a dilute suspension of the coating material, allowing a few seconds for the particles to settle, applying suction for about 30 sec, and then releasing the vacuum gently to avoid detachment of the thin coating. Radioactive suspensions are filtered and rinsed in a similar manner.

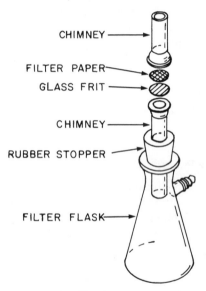

Figure 13-2   Apparatus for filtering a suspension of radioactive particles.

Samples filtered on flexible filter disks must be transferred to a firm support for counting. Planchets or cardboard squares, with the sample covered by a thin sheet of Mylar or Saran film, often are used. To prevent the filter disk from curling, the disc may be placed in a cupped planchet and held in position with a threaded or slit metal ring. Alternately, the filter disk may be positioned on a flat metal plate and held by an outer ring which slides over the plate.

Settling and Centrifugation of Slurries

A suspension of radioactive material sometimes is converted to a solid sample by centrifugation. After centrifugation, the cake of radioactive particles is washed and resuspended in a volatile solvent. This slurry is transferred to a cupped planchet for evaporation of the solvent and counting of the residue. Sample uniformity may be improved by tapping the planchet or by stirring the slurry gently during evaporation. Occasionally, samples of uniform thickness can be obtained by centrifuging the slurry onto a planchet mounted inside a specially designed centrifuge tube[16].

Electrodeposition

Very uniform films of radioactive samples, ranging from trace quantities to thicknesses of a few milligrams per square centimeter, often can be prepared by electrodeposition. This method is particularly useful for preparing α sources, and procedures have been described for electrodeposition of various α-emitting heavy elements[17, 18].

REFERENCES

[1] Walker, L., and R. Lougheed. 1965. A simple method for the assay of carbon-14 in compounds or mixtures. Intern. J. appl. Radiat. Isotopes 13:95.

[2] Belcher, E. 1960. The assay of tritium in biological material by wet oxidation with perchloric acid followed by liquid scintillation counting. Phys. Med. Biol. 5:49.

[3] Baxter, C., and I. Senoner. 1963. Liquid scintillation counting of ^{14}C-labeled amino acids on paper, using trinitrobenzene-1-sulfonic acid, and an improved combustion apparatus. Atomlight, No. 3, Nov. New England Nuclear Corp.

[4] MacDonald, A. 1961. The oxygen flask method. A review. Analyst 86:3.

[5] Oliverio, V., C. Denham, and J. Davidson. 1962. Oxygen flask combustion in determination of ^{14}C and ^{3}H in biological materials. Anal. Biochem. 4:188.

[6] Kelly, R. 1961. Determination of ^{14}C and ^{3}H in biological samples by Schöninger combustion and liquid scintillation techniques. Anal. Biochem. 2:267.

[7] Davidson, J., and V. Oliverio. 1968. Tritium and carbon-14 by oxygen flask combustion. In S. Rothchild (ed.), Advances in tracer methodology, Vol. 4. Plenum Press, New York. p. 67.

[8] Martin, L., and C. Harrison. 1962. The determination of ^{14}C- and tritium-labeled compounds in biological materials. Biochem. J. 82:18. (Proc. Biochem. Soc. 11 Nov. 1961, 410th meeting.)

[9] Sheppard, H., and W. Rodegker. 1962. Determination of ^3H and ^{14}C in biological materials using oxygen bomb combustion. Anal. Biochem. 4:246.

[10] Wilzbach, K., L. Kaplan, and W. Brown. 1953. The preparation of gas for assay of tritium in organic compounds. Science 118:522.

[11] Wilzbach, K., and W. Sykes. 1954. Determination of isotopic carbon in organic compounds. Science 120:494.

[12] Buchanan, D., and B. Corcoran. 1959. Sealed tube combustions for the determination of carbon-14 and total carbon. Anal. Chem. 31:1635.

[13] Peets, E., J. Florini, and D. Buyske. 1960. Tritium radioactivity determination of biological materials by a rapid dry combustion technique. Anal. Chem. 32:1465.

[14] Biggs, M., D. Dritchevsky, and M. Kirk. 1952. Assay of samples doubly labeled with radioactive hydrogen and carbon. Anal. Chem. 24:223.

[15] Isbell, H., and J. Mayer. 1959. Tritium-labeled compounds. II. General purpose apparatus, and procedures for the preparation, analysis, and use of tritium oxide and tritium-labeled lithium borohydride. J. Res. Nat. Bur. Standards 63A:177.

[16] Nuclear Chicago Technical Bulletin No. 7. How to prepare radioactive samples for counting on planchets, parts 1 and 2. Nuclear Chicago Corp., Des Plaines, Ill.

[17] Ko, R. 1957. Electrodeposition of the actinide elements. Nucleonics 15(1):72.

[18] O'Kelley, G. 1962. Detection and measurement of nuclear radiation, NAS-NS3105. Clearinghouse for Scientific and Technical Information. p. 127.

14

Sample Preparation for Liquid Scintillation Counting

A sample or "cocktail" prepared for liquid scintillation counting consists of at least three components: the radioactive material, an organic solvent, and an organic fluor. Additional components may be present to increase the solubility of the radioactive material, to enhance the transfer of energy from solvent to fluor molecules, or to match the fluorescence of the cocktail with the spectral sensitivity of the photomultiplier tubes used to measure the fluorescence. Since the radioactive material is mixed intimately with other components of the cocktail, radiation interactions in extraneous materials are reduced to a minimum; consequently, radiations with a very short range can be detected. For example, β particles from ^3H, ^{14}C, and ^{35}S can be detected with reasonable efficiency by liquid scintillation counting. The technique of liquid scintillation counting is outlined in Figure 14-1.

Solvents

A solvent for liquid scintillation counting should exhibit the following characteristics: (1) minimum absorption of light emitted by the fluor;

181

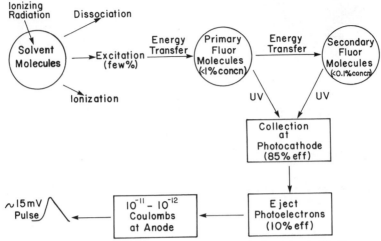

Figure 14-1 Formation of a voltage pulse in liquid scintillation counting.

(2) high absorption of energy from the radiation and transfer of this energy to the fluor; (3) maximum purity, to reduce interference with energy-transfer processes; (4) low freezing point, if the cocktail is to be counted at reduced temperature; and (5) high solubility for the radioactive material to be assayed. For a particular radioactive sample, the selection of a solvent usually requires a compromise among the solvent characteristics.

For many years, almost all solvents for liquid scintillation counting were aromatic hydrocarbons. Toluene still is used widely, particularly with lipids, fatty acids, steroids, and hydrocarbons, because it provides excellent energy transfer and is available at high purity and relatively low cost. Among the disadvantages of toluene are poor solubility for aqueous samples and a low flash point, making it somewhat hazardous to store and use. Other aromatic hydrocarbons which have been used as liquid scintillation solvents include benzene, xylene, ethylbenzene, and mesityl-ene.

Because of its miscibility with water, 1,4-dioxane is used widely as a solvent for aqueous radioactive samples. Energy is transferred less efficiently in 1,4-dioxane than in toluene, and counting efficiencies often are low for samples dissolved in 1,4-dioxane. The energy-transfer characteristics of 1,4-dioxane are improved greatly by the addition of naphthalene[1]. Because 1,4-dioxane freezes at 11°C, its use is restricted to temperatures above this value unless substances are added to depress the freezing point of the cocktail. Strongly quenching peroxides are formed in 1,4-dioxane upon standing, and the solvent should be purified immediately before use. Faissner, et al.[2], have reported that Shellsol A, a commercially available mixture of three trimethylbenzenes and other benzene derivatives, is a satisfactory solvent for many radioactive samples. This solvent exhibits energy-transfer characteristics comparable to those of toluene. Various solvents which have been evaluated for liquid scintillation applications are included in Table 14-1.

TABLE 14-1 *Relative Counting Efficiencies for ^{14}C-Labeled Samples in Various Solvents; All Solvents Contained 0.3% PPO as a Primary Fluor*[3, 4]

Solvent	Freezing point, °C	Relative efficiency
Toluene	−95	100
Methoxybenzene (anisole)	−37	100
Xylene (reagent, mixed isomers)	−20	97
1,3-Dimethoxybenzene	−52	81
n-Heptane	−90	70
1,4-Dioxane	+11	70
1,2-Dimethoxyethane (ethylene glycol dimethyl ether)	−71	60
Benzyl alcohol	−15	38
Diethylene glycol diethyl ether (diethyl carbitol)	−44	32
Acetone	−94	12
Tetrahydropyran	−81	6
Ethyl ether	−116	4
1,1-Diethoxyethane	−100	3
Tetrahydrofuran	−65	2
1,3-Dioxolane	−10	0
Ethyl alcohol	−114	0
Diethylene glycol monoethyl ether	−10	0
Ethylene glycol monomethyl ether	−85	0
Diethylene glycol	−8	0
Ethylene glycol	−13	0
2,5-Diethoxytetrahydrofuran	−27	0
n,n-Dimethylformamide	−61	0
Diethyl amine	−49	0
n-Methyl morpholine	−66	0
2-Ethylhexanoic acid	−117	0
Tri-*n*-butyl phosphate	−80	0

Fluors

A fluor for liquid scintillation counting should display the following characteristics: (1) high absorption of energy from solvent molecules and transfer of this energy to secondary fluor molecules or emission of light matching the spectral sensitivity of the photocathode; (2) solubility in solvent; (3) adequate chemical stability; (4) relatively low cost; and (5) short fluorescence decay time. Primary and secondary fluors employed most frequently in liquid scintillation counting are listed in Table 14-2. A secondary fluor should be added to the scintillation cocktail only if: (1) the cocktail contains a compound which interferes directly with light emission by the primary fluor; (2) the primary fluor is at a concentration which produces severe self-quenching; (3) the photocathode of the photo-multiplier tube responds more efficiently to light of longer wavelength; or (4) the cocktail contains a compound which absorbs light emitted by the primary fluor[5].

The spectral sensitivity of different photocathodes is compared in Figure 14-2 to the fluorescence spectra for various fluors. The concentration of fluor required for maximum fluorescence is illustrated in Figure

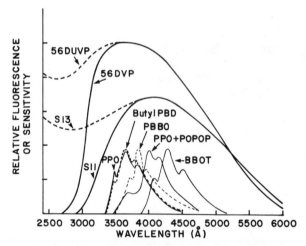

Figure 14-2 Spectral sensitivity of bialkali photocathodes (Philips photomultiplier tubes, type 56 DVP with Pyrex window and type 56 DUVP with quartz window) and Cs:Sb photocathodes (S-11 and S-13 response), compared to the fluorescence spectrum of PPO, butyl–PBD, PPO–POPOP, BBOT, and PBBO. [In part from J. Parmentier and E. ter Haaf. 1969. Developments in liquid scintillation counting since 1963. Internat. J. Appl. Radiat. Isotopes 20:305.]

TABLE 14-2 Various Fluors for Liquid Scintillation Counting

Fluor	Abbreviation	Type	Relative photon Yield[4]	Wavelength of maximum fluorescence, Å
p-Terphenyl	TP	Primary	1.33	3406
2,5-Diphenyloxazole	PPO	Primary	1.01	3800
2-Phenyl-5-(4-biphenylyl)-1,3,4-oxadiazole	PBD	Primary	1.24	3610
2-(4'-t-Butylphenyl)-5-(4"-biphenyl)-1,3,4-oxadiazole	Butyl-PBD	Primary	~1.20	3660
2,5-Bis-2-(5-t-butylbenzoxazolyl)thiophene	BBOT	Primary	~1.10	4350
2,5-Diphenyl-1,3,4-oxadiazole	PPD	Primary	1.24	3600
2-(4'-Biphenylyl)-6-phenylbenzoxazole	PBBO	Primary	1.25	3960
1,4-Bis-2-(5-phenyloxazolyl)benzene	POPOP	Secondary	1.17	4300
1,4-Bis-2-(4-methyl-5-phenyloxazolyl)benzene	Dimethyl–POPOP	Secondary		4300
1,4-Di-(2-tolyloxazolyl) benzene	TOPOT	Secondary		4380

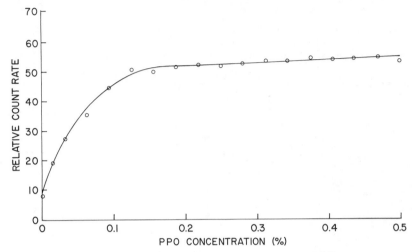

Figure 14-3   Relative fluorescence as a function of PPO concentration in toluene.

14-3 for PPO in toluene. From the data in Figure 14-3, a concentration of 4 g of PPO per liter of toluene might be chosen. Fluors should not be used at concentrations greater than those necessary for maximum fluorescence, because most fluors are costly and some act as quench agents at higher concentrations, thereby reducing the fluorescence of the cocktail.

Scintillation Cocktails

A large number of scintillation cocktails have been proposed with a variety of primary and secondary fluors, solvents, and additives. However, only the small number of cocktails which are used widely are discussed here.

Perhaps the scintillation cocktail used most often is toluene with 0.3–0.5% PPO and, when necessary, 0.01–0.03% POPOP. This solution is useful particularly for radioactive materials soluble in toluene, for samples counted in suspension, and for paper chromatograms. Other materials may be counted at reasonable efficiency in toluene–PPO by adding agents to the cocktail to improve sample solubility. For example, ethanol, methanol, 1,4-dioxane, and 2-ethoxyethanol have been added to toluene–PPO to increase the solubility of aqueous compounds.

Aqueous samples most often are counted in dioxane–PPO (plus POPOP or dimethyl POPOP, if necessary) to which naphthalene has

TABLE 14-3 Selected Cocktails for Liquid Scintillation Counting[10]

| Solvent | | Fluor | | Figure of merit |
Principal	Intermediate	Primary, g/l	Secondary, mg/l	
Toluene		PPO, 3–5	POPOP, 50–100 Dimethyl POPOP, 100–200	
Toluene		p-Terphenyl, 5	Dimethyl POPOP, 500	4.6–16.2
Toluene + Ethanol		PPO	POPOP	32.3
Xylene (5 parts) + dioxane (5 parts) + ethanol (3 parts)	Naphthalene, 80 g/l	PPO, 5	α-NPO 50	
Xylene (1 part) + dioxane (3 parts) + cellosolve (3 parts)	Naphthalene, 80 g/l	PPO, 10	POPOP 500	~ 60
Dioxane		p-Terphenyl, 5		11.4
Dioxane	Naphthalene, 20–120 g/l	PPO 4–10	POPOP, 50–300	100–200
Dioxane + methanol, ethylene glycol	Naphthalene, 60 g/l	PPO, 4	POPOP, 200	100
Dioxane (6 parts) + 1,2-dimethoxy-ethane (1 part)	Anisole, 1 part	PPO, 12	POPOP, 50	50–80
1,2-dimethoxyethane	Naphthalene, 100 g/l	PPO, 7	POPOP, 300	40–45
Dioxane (2 parts) + 1,2-dimethoxy-ethane (1 part)	Naphthalene, 100 g/l	PPO, 7	POPOP, 300	75–80

been added at a concentration of 20–120 g/liter. The solution employed most frequently contains 100 g of naphthalene, 7–10 g of PPO and, if required, 300 mg of POPOP per liter of dioxane. This mixture is useful for concentrations of water up to 20%, at which counting efficiencies greater than 10% can be obtained for ^{3}H[6].

A counting solution known as Bray's solution contains 60 g of naphthalene, 4 g of PPO, 200 mg of POPOP, 100 ml of methanol, 20 ml of ethylene glycol and enough dioxane to make 1 liter of solution[7]. This solution exhibits good solubility for aqueous samples and relatively high (for example > 10%) counting efficiencies for ^{3}H.

Another cocktail of limited popularity is the "611 cocktail," composed of 6 parts dioxane, 1 part anisole, 1 part dimethoxyethane, and 12 g/liter PPO. Chemiluminescence with this cocktail can be suppressed by adding a trace of an antioxidant such as di-*t*-butyl-4-hydroxytoluene which suppresses the formation of ether peroxides[6]. A cocktail containing 1 vol of Triton-X-100* to 2 vol of (toluene + 0.5% PPO + 0.01% POPOP) has been described[8, 9] as acceptable for tritiated water and a variety of polar and nonpolar organic compounds.

The acceptability of a scintillation cocktail sometimes is described by its figure of merit, defined as the product of the counting efficiency and the concentration of sample in the cocktail. For example, if a 10% concentration of ^{3}H-water in a cocktail is counted at an efficiency of 10%, then the figure of merit of the cocktail is 10×10 or 100 for ^{3}H-water.

Solubilization and Decolorization of Samples

Biologic tissues are difficult to dissolve into solution. Consequently, the usefulness of liquid scintillation counting would be limited if methods were not available to improve the solubility of biologic samples in various cocktails. Most of these methods involve heating of the sample under strongly basic conditions in an effort to degrade polymeric materials to soluble products of lower molecular weight. A few solubilizing agents and some reported applications are described in Table 14-4[11, 12].

Hyamine 10-X* (*p*-diisobutylcresoxyethoxyethyl dimethylbenzyl ammonium chloride) is a germicidal quaternary converted to hydroxide form by a technique such as addition of silver oxide to a methanolic solution of pure material. Proteins solubilized in Hyamine 10-X often produce chemiluminescence which usually can be suppressed by acidification and dark adaptation of the samples. Hyamine 10-X is a severe quench agent, and should not be used in quantities greater than absolutely necessary to achieve complete solubilization.

*Trademark of Rohm and Haas, Inc.

TABLE 14-4 *Selected Solubilizing Agents and Applications*

Biologic material	Solubilizing agent
Whole tissue, tissue extracts, biologic fluids, proteins, amino acids	NCS (mixture of organic quaternary ammonium bases)
Bacterial cells	Formamide
Dry tissue, protein, fresh tissue, serum	Potassium hydroxide/Hyamine 10-X chloride
Plasma, urine	Aqueous potassium hydroxide
Tissue	2 N Methanolic potassium hydroxide
Proteins and amino acids, plasma, blood	Hyamine 10-X hydroxide
Nucleic acids	Hydrochloric acid/Hyamine 10-X hydroxide
Protein	88% Formic acid
Blood, tissue	60% Perchloric acid
Fluids, tissue, bone	30% Hydrogen peroxide

The solubilizing agent NCS† is a mixture of bases with molecular weights ranging from 250 to 600. For a wide variety of biologic materials, the solubilizing properties of NCS are comparable to those of Hyamine 10-X[13].

The effectiveness of most solubilizing agents is reduced by dilution. Consequently, samples should be dissolved in solubilizer before being added to the scintillation cocktail. For some samples, ultrasonic agitation may facilitate the solubilization process[14].

Biologic samples which yield colored solutions when solubilized may produce severe quenching and low counting efficiencies. These problems can be resolved by combustion or oxidation of the samples, with capture of the radioactive products and their incorporation into a scintillation cocktail. For example, tritiated water can be added directly to dioxane or Bray's solution, and $^{14}CO_2$ and $^{35}SO_2$ may be captured in an alkaline absorbent (for example, phenylethylamine, ethanolamine, Primene-81-R,* Hyamine 10-X, or potassium hydroxide) which can be added to the scintillation solution. Combustion or oxidation of samples is tedious and sometimes unnecessary, provided that bleaching agent is available to remove the color from the sample or scintillation cocktail. Bleaching agents which have been used include hydrogen peroxide, chlorine water, benzoyl peroxide, sodium borohydride, methyl trioctyl, and ammonium borohydride[11].

---

†Trademark of Amersham–Searle Division, Nuclear-Chicago Corporation.
*Trademark of Rohm and Haas, Inc.

Suspension and Emulsion Counting

Samples which are difficult to dissolve in a scintillation cocktail sometimes may be counted as suspensions. Just before counting, each sample
should be agitated to minimize the change in count rate which occurs as
the suspended particles settle to the bottom of the vial. The suspended
particles interfere with energy-transfer processes in the cocktail and also
absorb light emitted by the fluor. Consequently, samples counted in
suspension provide less light for the photocathodes. The interference of
the particles with energy-transfer processes can be reduced by counting
the sample as a suspension of very fine particles. However, this procedure expands the total surface area of the particles and increases the
loss of light due to multiple internal reflection. Interference with energy
transfer and internal reflection of light tend to balance each other as
particle size is varied, and most samples are prepared for suspension
counting with little concern for particle size.

The settling of particles suspended in a cocktail can be suppressed by
increasing the viscosity of the cocktail. Aluminum stearate was used
originally to thicken toluene solutions[15], and Thixcin-R, a derivative of
castor oil, was introduced later for this purpose[16]. The gelling agent
now used most often is Cab-O-Sil,* an aerated silica with an average
particle size of 15 μ and a surface area of 200 m²/g. This thixotropic agent
(fluid when shaken, firm gel at rest) can support up to 2 g of $BaCO_3$ when
added to a 20 ml volume of toluene–PPO–POPOP at a concentration
of about 4% by weight.

Measurement of $^{14}CO_2$ precipitated as $BaCO_3$ or trapped in NaOH
solution is one of the more common applications of suspension counting.
Other applications include the counting of NaOH digests of tissue, ion
exchange resins, inorganic salts, lyophilized tissue and bacteria, and
scrapings from thin-layer chromatograms[17].

Suspension counting techniques are limited somewhat by reduced
counting efficiencies and by difficulties in determining the counting
efficiencies for samples in suspension. These difficulties are caused
primarily by variations in the interference of the particles with energy-
transfer processes. The problem of reduced counting efficiency is particularly severe for weak β particles from tritium, and suspension counting
of tritiated samples usually is not satisfactory. For samples emitting β
particles of higher energy, the counting efficiency may be estimated by
counting a sample of known activity as a suspension with a particle-size
distribution similar to that for the samples. Usually, this procedure
involves laborious techniques for grinding and sieving the standard

*Trademark of the Cabot Corporation.

sample. Samples of barium carbonate are excluded from this requirement because the counting efficiency of barium carbonate does not vary with particle size[11].

Undissolved samples sometimes can be counted as a true emulsion rather than as a suspension. The detergent Triton-X-100, when added to toluene–PPO at a concentration of 1 part detergent to 2 parts toluene–PPO by volume, provides a medium for counting a variety of aqueous samples at reasonable efficiencies. For example, counting efficiencies for ^{14}C-labeled samples in emulsion are comparable to those achieved for homogeneous solutions. The counting efficiencies for samples of ^{3}H-water in Triton-X-100:toluene have been reported to be reproducible but lower than those for homogeneous cocktails containing tritium. The preparation of samples for emulsion counting is relatively simple, and this technique is increasing in popularity.

Samples on Paper and Filter Disks

Radioactive material attached to a paper chromatogram or disk of filter paper may be counted by immersing the chromatographic strip or filter disk in a vial of scintillation solution. If the radioactive material is insoluble in the scintillation solution, then this procedure may provide reproducible results; however, if the material passes from the strip or disk into solution, then the count rate may vary significantly with time. If the radioactivity migrates slowly into solution, then the sample should be removed completely from the paper and counted in solution or as a combustion product. For samples attached to paper chromatograms or filter paper, the counting efficiency should be determined by internal standardization; other methods for estimating counting efficiency sometimes provide erroneous results[18].

For most β-emitting samples, the orientation of the chromatographic strip or filter disk in the counting vial does not greatly affect the counting efficiency. Samples containing tritium may be an exception, however, because the photocathodes are shielded from light by varying amounts with the paper in different orientations, and the counting efficiency changes rapidly with the amount of light reaching the photocathodes. To achieve reproducible results, some investigators always place paper containing tritium on the bottom of the counting vial, even though the counting efficiency is reduced significantly with the paper in this position. Newer liquid scintillation units with pulse summation reduce the dependence of counting efficiency upon the orientation of a tritium sample in the counting vial.

Radioactive samples insoluble in toluene have been counted by evaporating the samples on filter paper and immersing the paper in scintillation solution. Potassium gluconate derived from blood glucose, amino acids, protein hydrolysates, proteins, and nucleotides have been counted in this manner[19, 20]. Nakshbandi has suggested that paper strips impregnated with the plastic scintillator Naton 136 provide counting efficiencies for tritium higher than those achieved by more conventional methods[21]. Filter paper moistened with an alkaline trapping agent (such as aqueous KOH or Hyamine) can be used to collect and count small quantities of $^{14}CO_2$ released during biologic oxidation.

Glass-fiber filters sometimes are preferable to filter paper, particularly when smaller tritium-labeled molecules penetrate the filter paper by varying amounts from sample to sample. The variable penetration, which can cause differences in counting efficiency from one sample to the next, can be reduced by using glass-fiber disks comprised of a compressed mat of tiny glass threads. A glass-fiber disk can be counted intact by one of the methods outlined above, or in suspension by using vigorous agitation to break the disk in the counting vial.

Millipore filters, consisting of a nonporous matrix of plastic which is punctured with holes of uniform diameter, are soluble in dioxane. A dioxane-soluble sample can be collected on a Millipore filter and counted in solution by dissolving the sample and filter in dioxane.

Suspended Scintillators

A homogeneous solution of radioactive sample, solvent, fluor, and additives is the type of scintillation cocktail which usually furnishes highest counting efficiency and greatest reproducibility. When a homogeneous solution cannot be achieved, the sample usually is counted as a suspension or emulsion in a solvent-fluor solution. Occasionally, however, a cocktail is useful which contains the fluor in suspension and the sample in solution or in gaseous form. For example, ^{14}C in aqueous and alcoholic solutions can be detected by packing solid scintillation filaments in a counting vial and covering the filaments with the sample dissolved in solvent[22]. The filaments can be reused after rinsing, provided that all residual activity has been removed.

Although many solid scintillators are available, crystalline anthracene (blue-violet fluorescence grade) is probably the scintillator most suitable for most applications. A trace of a wetting agent (such as Triton GR-5, 1:1000) usually is added to the counting vial to ensure intimate contact between the solution and the solid scintillator. Counting efficiencies are listed in Table 14-5 for a few nuclides assayed with anthracene crystals [23].

TABLE 14-5 *Counting Efficiencies For Nuclides Counted with Suspended Anthracene Crystals*

Nuclide	^3H	^{14}C	^{45}Ca	^{32}P	^{131}I
Efficiency, %	0.5	16–20	49	78	58

Elimination of chemical conversion and preliminary handling of the sample, permitting the sample to be counted immediately upon receipt, are the major advantages of using suspended scintillators. Often, the sample can be recovered after assay. The continuous monitoring of effluents from column chromatographs is perhaps the most widespread use of suspended scintillators.

Continuous Flow Monitors

The development of counting techniques with suspended scintillators has contributed greatly to the design of flow counters for monitoring the radioactivity in gaseous and liquid effluents from chromatographs. Most flow counters utilize anthracene crystals packed in a flow cell similar to that illustrated in Figure 14-4. In this counter, the flow cell is positioned between plastic light pipes separating two photomultiplier tubes. Using a flow counter of this type, Rapkin and Gibbs[24] achieved counting efficiencies greater than 50% for gaseous and aqueous ^{14}C activity, about 2% for aqueous ^3H activity, and approximately 7% for gaseous ^3H. Residual activity was not detectable on the anthracene crystals, and the flow counter did not degrade the resolution of data from amino acid analyzers. The background count rate for this counter can be reduced by using a transparent plastic such as Lucite in place of glass for construction of the counter.

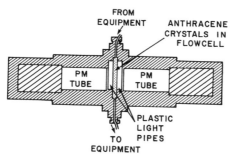

Figure 14-4 Typical anthracene flow cell used to monitor radioactivity in liquid or gaseous streams[24].

A flow counter developed by Schram and Lombaert[25] consists of a 60 cm length of polyethylene tubing (2.2 mm inside diameter) wound into a flat coil (Figure 14-5). The coil is filled with anthracene crystals and positioned in a Lucite holder containing silicone oil for optical coupling of the coil and holder. The assembly is positioned between two photo-multiplier tubes mounted with springs to provide optical contact. With anthracene crystals 150 μ in diameter in a coil-type counter, Schram and Lombaert achieved counting efficiencies of 55% for ^{14}C and 2% for [3]H. For monitoring chromatographic effluents which react chemically with anthracene, crystals of *p*-terphenyl coated with silicone often can be used in place of anthracene[26].

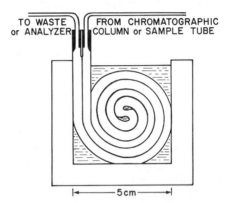

Figure 14-5   Coil-type anthracene flow cell.

Another approach to continuous flow monitoring is the use of a scintillation solution in place of suspended scintillators[27]. The chromatographic effluent is trapped in a scintillation solution circulated continuously between two photomultiplier tubes. This type of counter often provides counting efficiencies for ^{3}H and ^{14}C which are higher than those obtainable with flow cells containing suspended scintillators. Also, flow counters containing scintillation solution are less subject to radioactive contamination. Disadvantages of this type of counter include:

(1) Poor solubility of many chromatographic effluents in scintillation solutions
(2) Problems with salts present in the effluent
(3) Cost of large volumes of scintillation solution which can be used only once
(4) Variations in counting efficiency with composition of the effluent
(5) Destruction of the sample

Flow counters have been designed with flow cells machined from a block of solid scintillator. For example, Schram and Lombaert[28] have developed a flow cell in which the radioactive stream flows through a spiral chamber in a block of plastic scintillator. Cells made of small-bore tubing of plastic scintillator which is wound into a flat coil have been used by Kimbel and Willenbrink[29] and by Funt and Heterington[30] to achieve counting efficiencies greater than 75% for ^{32}P, 50% for ^{22}Na, and 5% for ^{14}C.

Cerenkov Counting

As a charged particle traverses a medium, molecules of the medium are polarized along the path of the particle. Almost immediately, these molecules return to their quiescent state by emission of electromagnetic radiation. If the velocity of the particle is less than the velocity of light in the medium, then the radiation from the individual molecules interferes destructively, and none is released from the medium. However, if the particle's velocity exceeds the velocity of light in the medium, then some of the radiation interferes constructively and escapes from the medium. This radiation falls in the visible and near-ultraviolet region of the electromagnetic spectrum and is known as Cerenkov radiation.

A few investigators have described the assay of β-emitting samples by detection of Cerenkov radiation rather than the light emitted by a fluor. For a sample to be counted by detection of Cerenkov radiation, the maximum energy E_{max} of the β particles from the sample must exceed

$$E_{max} > 0.511 \left(\frac{1}{\sqrt{1 - (1/n^2)}} - 1 \right) \text{MeV}$$

where n is the index of refraction of the medium containing the sample. A solution which contains a sample emitting β particles above this threshold energy may be counted simply by placing the solution in a counting vial between two photomultiplier tubes. The efficiency for Cerenkov counting increases rapidly with increasing β-particle energy above the threshold energy (Figure 14-6). For some applications of Cerenkov counting, a wavelength shifter such as sodium potassium salt of 2-naphthylamine-6, 8-disulphonic acid (100 mg/l)[31], or 7-amino-1, 3-naphthalene disulphonic acid (ANDA)[32], may be required to absorb some of the ultraviolet Cerenkov radiation and to reemit the energy as visible light.

Reported applications for Cerenkov counting include the measurement of ^{90}Sr in urine following radiation accidents[33], the use of ^{42}K in the determination of exchangeable potassium in urine[34], the use of ^{24}Na

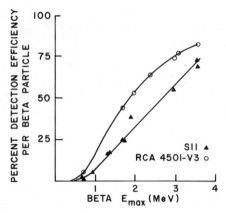

Figure 14-6 Percent detection efficiency per β particle for Cerenkov counting with a photomultiplier tube with an S-11 response and with an RCA 4501-V3 photomultiplier tube containing a high quantum efficiency photocathode. [From R. Parker, and E. Elrick. 1970. Cerenkov counting as a means of assaying β-emitting radionuclides. *In* E. Bransome, Jr. (ed.), The current status of liquid scintillation counting, Grune and Stratton, New York. p. 111.]

to measure steroid hormone uptake in tissue slices[35], and the study of ion transport of plants with ^{86}Rb[32].

Counting Vial

Vials used in liquid scintillation counting usually are made of glass or polyethylene and will hold about 20 ml of scintillation cocktail. Glass vials usually are composed of special glass with low potassium content to reduce the contribution of naturally occurring ^{40}K to the background count rate. Polyethylene vials contain no potassium, and the background count rate may be lower if these vials are used in place of glass vials; however, some polyethylene vials have been reported to be permeable to toluene[36].

REFERENCES

[1] Kallman, H., M. Furst, and F. Brown. 1962. Scintillation solution enhancers, U.S. Patent 3,068,178, Dec.

[2] Faissner, H., F. Ferrero, A. Ghani, and M. Reinharz. 1963. New scintillation liquids. Nucleonics 21(2):50.

[3] Davidson, J., and P. Feigelson. 1957. Practical aspects of internal-sample liquid scintillation counting. Intern. J. Appl. Radiat. Isotopes 2:1.

[4] Kowalski, E., R. Anliker, and K. Schmid. 1967. Criteria for the selection of solvents in liquid scintillation counting: New efficient solvents with high solubility. Intern. J. Appl. Radiat. Isotopes 18:307.

[5] Bush, E., and D. Hansen. 1965. Radioisotope sample measurement techniques in medicine and biology. IAEA, Vienna. p. 395.

[6] Rapkin, E. 1967. Preparation of samples for liquid scintillation counting. In G. Hine (ed.), Instrumentation in nuclear medicine. Academic Press, New York. p. 181.

[7] Bray, G. 1960. A simple efficient liquid scintillator for counting aqueous solutions in a liquid scintillation counter. Anal. Biochem. 1:279.

[8] Turner, J. 1969. Tritium counting with the Triton-X-100 scintillant. Internat. J. Appl. Radiat. & Isotopes 20:499.

[9] Patterson, M., and R. Greene. 1965. Measurement of low-energy β-emitters in aqueous solution by liquid scintillation counting of emulsions, Anal. Chem. 37:854.

[10] Raplin, E. 1967. Preparation of samples for liquid scintillation counting, part 1. Picker Laboratory Scintillator 11(3L), Jan.

[11] Turner, J. Sample preparation for liquid scintillation counting, Radiochemical Reviews, Amersham–Searle Div., Nuclear-Chicago Corp.

[12] Rapkin, E. 1967. Preparation of samples for liquid scintillation counting, part II. Picker Laboratory Scintillator 11(4L), Feb.

[13] Hansen, D., and E. Bush. 1967. Improved solubilization procedures for liquid scintillation counting of biological materials. Anal. Biochem. 18:320.

[14] Bruno, G., and J. Christian. 1960. Note on suitable solvent systems usable in the liquid scintillation counting of animal tissue. J. Am. Pharm. Assoc., Sci. Ed. 49:560.

[15] Funt, B. 1956. Scintillating gels. Nucleonics 14(8):83.

[16] White, C. G., and S. Helf. 1956. Suspension counting in scintillating gels. Nucleonics 14(10):46.

[17] Rapkin, E. 1967. Preparation of samples for liquid scintillation counting, part III. Picker Laboratory Scintillator 11(5L), March.

[18] Furlong, N. 1970. Liquid scintillation counting of samples on solid supports. In E. Bransome, Jr. (ed.), The current status of liquid scintillation counting. Grune and Stratton, New York. p. 201.

[19] Blair, A., and S. Segal. 1962. Use of filter paper mounting for determination of the specific activity of gluconate-^{14}C by liquid scintillation assay. Anal. Biochem. 3:221.

[20] Davies, J., and E. Cocking. 1966. Liquid scintillation counting of ^{14}C and ^{3}H samples using glass-fibre or filter-paper disks. Biochim. Biophys. Acta 115:511.

[21] Nakshbandi, M. 1965. A plastic scintillator method for the radioassay of ^{3}H- and ^{14}C-labeled compounds on filter paper. Internat. J. Appl. Radiat. Isotopes 16:157.

[22] Steinberg, D. 1959. Radioassay of aqueous solutions mixed with solid crystalline fluors. Nature 183:1253.

[23] Steinberg, D. 1963. A new technique for counting aqueous solutions in the liquid scintillation spectrometer. *In* S. Rothchild (ed.), Advances in tracer methodology, vol. 1. Plenum Press, New York. p. 93.

[24] Rapkin, E., and J. Gibbs. 1962. A system for continuous measurement of radioactivity in flowing streams. Nature 194:34.

[25] Schram, E., and R. Lombaert. 1962. Determination of tritium and carbon-14 in aqueous solution with anthracene powder. Anal. Biochem. 3:68.

[26] Karmen, A. 1963. Measurement of tritium in the effluent of a gas chromatography column. Anal. Chem. 35:536.

[27] Hunt, J. 1968. Continuous flow monitor system for detection of UV absorbance, ^{14}C, and ^{3}H in effluent of a column chromatogram. Anal. Biochem. 23:289.

[28] Schram, E., and R. Lombaert. 1957. Determination continue du carbone-14 et du soufre-35 en milieu Aquenx par un dispositif au scintillation. Application aux effluents chromatographiques. Anal. Clin. Acta 17:417.

[29] Kimbel, K., and J. Willenbrink. 1958. Fortlaufende, Messang schwacher β-Strahlen in Flussigkeiten mit Szintellatorschlauch. Naturwissenschaften 45:567.

[30] Funt, B., and A. Hetherington. 1959. Spinal capillary plastic scintillation flow counter for beta assay. Science 129:1429.

[31] Elrick, R., and R. Parker. 1966. The assay of β-emitting radioisotopes using Cerenkov counting, Internat. J. Appl. Radiat. Isotopes 17:361.

[32] Lauchli, A. 1969. Radioassay for β-emitters in biological materials using Cerenkov radiation, Internat. J. Appl. Radiat. Isotopes 20:265.

[33] Narrog, J. 1965. Detection of beta-emitting nuclides of energy > 1 MeV in urine after an accident by means of measurement of the Cerenkov effect in a liquid scintillation counting system. *In* Personnel dosimetry for radiation accidents. IAEA, Vienna. p. 427.

[34] Francois, B. 1967. La detection du potassium-42 par l'effect Cerenkov, Internat. J. Appl. Radiat. Isotopes 18:525.

[35] Braunsberg, H., and A. Guyver. 1965. Automatic liquid scintillation counting of high-energy β emitters in tissue slices and aqueous solutions in the absence of organic scintillator. Anal. Biochem. 10:86.

[36] Rapkin, E., and J. Gibbs. 1963. Polyethylene containers for liquid scintillation spectrometry. Internat. J. Appl. Radiat. Isotopes 14:71.

15

Liquid Scintillation Counting

In the earliest liquid scintillation counters, a transparent container containing the sample was coupled optically to a single photomultiplier tube. The container was surrounded by reflectors to increase the collection of light by the photocathode. The signal from the photomultiplier tube was directed through a preamplifier, amplifier, and pulse-height analyzer and recorded by a scaler (Figure 15-1). To reduce electronic noise which masked the signal at room temperature, photomultiplier tube and preamplifier were operated at reduced temperature (0–5°C).

Noise was reduced further by adding a second photomultiplier tube and associated circuits, including a coincidence circuit, to the counter (Figure 15-2). The coincidence circuit transmits a pulse to the scaler

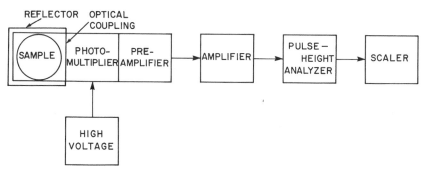

Figure 15-1 A liquid scintillation counter with a single photomultiplier tube.

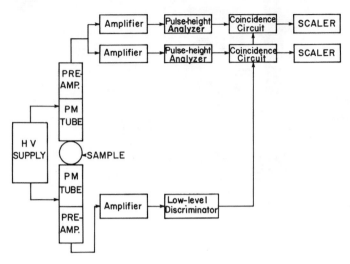

Figure 15-2 A liquid scintillation counter with two photomultiplier tubes and a coincidence circuit to reject noise pulses.

only when a pulse is received simultaneously (within the resolving time of the coincidence circuit) at both input terminals of the circuit. Noise pulses originating in one photomultiplier tube have no corresponding pulse in the second circuit, and these pulses usually are rejected by the coincidence circuit and are not recorded by the scaler. Noise pulses are recorded only when they happen to arrive simultaneously at the input terminals of the coincidence circuit. In the circuit in Figure 15-2, two separate channels are available for simultaneous counting of pulses in two separate pulse-height ranges. Counters of this design were sold widely until about 1963.

In addition to two photomultiplier tubes and a coincidence circuit, recent liquid scintillation counters contain a pulse-summation circuit. This circuit increases the amplitude of the signals representing radiation interactions by adding the signals from the two photomultiplier tubes. Pulse summation also provides a more narrow and reproducible pulse-height spectrum for the β particles being counted. A block diagram of a modern liquid scintillation counter is shown in Figure 15-3.

Light Collection and Detection

In a liquid scintillation cocktail containing a β-emitting isotope, approximately seven photons of light are released for every kiloelectron volt of

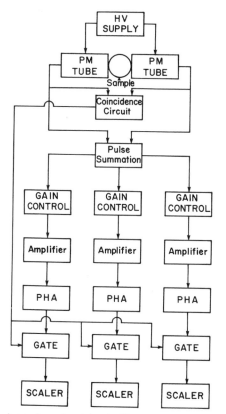

Figure 15-3 A modern liquid scintillation counter with a coincidence circuit and pulse summation.

energy absorbed. For example, the average energy is 5.6 keV for β particles from ^{3}H, and about 40 light photons are released for each decay of ^{3}H in an unquenched cocktail. This small number of photons is reduced by quenching and by light losses in the coupling mechanism between counting vial and photomultiplier tube. In a liquid scintillation counter with two photomultiplier tubes, half of the remaining light photons impinge upon each photocathode. The quantum efficiency (number of electrons released per incident photon) is about 0.2–0.3 for most photocathodes; consequently, 3–4 electrons are released from each photocathode on the average for each decay of ^{3}H in the cocktail. This number of electrons is increased by 10^5–10^8 in the dynode section of the photomultiplier tube. The signal from the photomultiplier tube is transmitted to a pulse-summation circuit (Figure 15-3). By summing the signals from both photomultiplier tubes, the pulse-summation circuit effectively doubles the amplitude of the signals representing radiation interactions, and permits better discrimination between true signals and electronic noise generated in the photomultiplier tubes. With a pulse-summation circuit, statistical fluctuations in the number of light photons reaching each photocathode produce less distortion in the pulse-height spectrum. Consequently, this circuit furnishes a pulse-height spectrum which resembles more closely the energy spectrum of the β particles being counted. Pulse-height spectra for ^{3}H obtained with and without pulse-height summation are shown in Figure 15-4. From the summation circuit, the signal is transmitted to a series of channels (three in Figure 15-3) consisting of a linear amplifier, a pulse-height analyzer, and a display device such as a scaler.

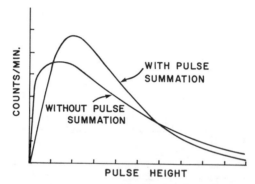

Figure 15-4 Pulse-height spectra for ^{3}H obtained with and without pulse summation. The spectrum without pulse summation was obtained with an amplifier gain twice that for the pulse-summation spectrum. [From E. Polic. 1967. Liquid scintillation counting equipment. *In* G. Hine (ed.): Instrumentation in nuclear medicine. Academic Press, New York. p. 239.]

In newer liquid scintillation counters, the use of photomultiplier tubes with as many as 13 dynodes eliminates the requirement for pre-amplifiers between the photomultiplier tubes and the pulse summation circuit. Without preamplifiers present to lengthen the pulses, very narrow pulses (short rise and decay time) may be transmitted to the coincidence circuit. With these narrow pulses, a coincidence circuit with a small resolving time (10–30 nsec) is used to provide better discrimination between true pulses and electronic noise. The number of "accidentals" (noise pulses transmitted by the coincidence circuit) may be estimated with the expression

$$\text{accidentals (cpm)} = 2c_1c_2\tau$$

where c_1 = noise rate in cpm from one photomultiplier tube, c_2 = noise rate in cpm from the second photomultiplier tube, and τ = the resolving time of the coincidence circuit in minutes[1].

Example 15-1

What is the contribution of accidentals to the count rate in a liquid scintillation counter with photomultiplier tubes contributing a noise rate of 30,000 cpm and with a coincidence circuit resolving time of 30 nsec?

$$\text{accidentals (cpm)} = 2c_1c_2\tau$$

$$= 2(3 \times 10^4 \text{ cpm})(3 \times 10^4 \text{ cpm})\frac{(30 \times 10^{-9} \text{ sec})}{60 \text{ sec/min}}$$

$$\simeq 1 \text{ cpm}$$

Noise rates have been reduced significantly in newer photomultiplier tubes with bialkali photocathodes containing potassium in addition to cesium and antimony. With these tubes and a coincidence circuit with a resolving time of 10 nsec, the accidental count rate can be reduced to 0.1 cpm.

To prevent extraneous light from striking the photocathodes of the photomultiplier tubes, the counting chamber for the sample vial and the housing for the photomultiplier tubes must be impermeable to light. These components also should be shielded from cosmic radiation and radiation from nearby sources. Samples should be positioned in a rapid and reproducible manner in the counting chamber, and the transfer of light from the vial to the photocathodes should be as efficient as possible. In most liquid scintillation counters, the sample vial is positioned within a short light guide which is coupled optically to the photomultiplier tubes. The vial should enter the light guide with adequate clearance to prevent fluorescence induced by friction, and the light guide should be coated with a diffuse reflecting material (such as titanium dioxide) to increase the collection of light by the photocathodes. The production of Cerenkov radiation by β particles emitted by ^{40}K and uranium isotopes in the glass envelopes of the photomultiplier tubes can be minimized by using photomultiplier tubes with faces of quartz rather than glass.

In a liquid scintillation counter employing pulse summation, the amplification must be identical for both photomultiplier tubes. This requirement necessitates a separate high-voltage adjustment for each tube. To maintain the correct high voltage on the photomultiplier tubes, the high-voltage supplies for the tubes must be exceptionally stable.

Pulse-Height Analysis

From the summation circuit, voltage pulses are transmitted to 1–5 data channels, each containing an amplifier and a pulse-height analyzer. In each channel, the gain of the amplifier and the window of the analyzer can be varied to select pulses in a particular range of pulse heights. Pulses

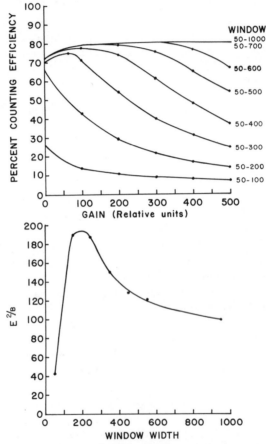

Figure 15-5 Determination of settings for balance-point counting of [14]C. Top, family of curves of counting efficiency for a ^{14}C sample as a function of amplifier gain for analyzer windows of different widths. For each window width, the optimum gain is that which provides the greatest counting efficiency. Bottom, plot of E^2/B at optimum gain settings as a function of window width for the data in the figure above. The optimum window width is that which provides the greatest value for E^2/B (200 divisions or 50–250).

selected by the analyzer are transmitted to the display device through an electronic gate which is controlled by the coincidence circuit and is opened for true pulses but closed for pulses caused by electronic noise. Usually, the window of the analyzer and the gain of the amplifier are adjusted for balance-point counting, defined as the window and gain settings which provide a maximum value of (counting efficiency)^2/ (background count rate) for detection of the radiation from a particular nuclide. The determination of settings for balance-point counting for ^{14}C is illustrated in Figure 15-5. When a sample is counted under balance-point conditions, slight variations in signal size caused by quenching or electronic instability do not greatly influence the count rate, because the efficiency remains relatively constant with slight shifts in gain (Figure 15-5). Newer liquid scintillation counters offer improved electronic stability and may not require exact determination of balance point settings. However, balance-point counting remains very useful when variable quenching (discussed later in this chapter) is encountered from one sample to the next[2]. Also, for a sample of low activity, the precision of the net count rate is greatest if the sample is counted under balance-point conditions.

The influence of amplifier gain on the pulse-height spectrum for ^{3}H is illustrated in Figure 15-6. As the gain increases, the spectrum encompasses a greater range of pulse heights, and the amplitude of the spectrum

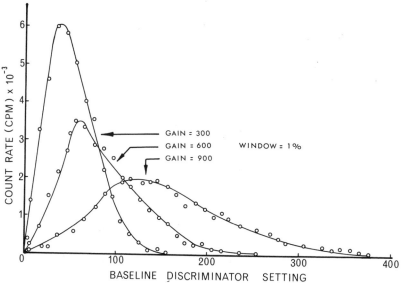

Figure 15-6 The influence of amplifier gain on the pulse-height spectrum for ^{3}H.

decreases at each pulse-height setting. At higher settings of amplifier gain, the spectrum is essentially flat, and variations in amplification or quenching have little effect on the count rate. By counting a sample at high gain, the influence of variable quenching and unstable electronics on counting efficiency can be minimized. At these settings, a loss of counts from pulses shifted from the window to positions below the lower discriminator is compensated by an equal number of counts shifted into the window from above the upper discriminator. This approach to sample counting is referred to as *flat spectrum operation*, and is useful only for high-activity samples which provide adequate count rates at very low counting efficiencies.

Quenching

Quenching refers to any process causing a reduction in the amount of light incident upon the photocathodes of a liquid scintillation counter. At least three quenching mechanisms can be identified; these mechanisms are chemical quenching, color quenching, and optical quenching.

Chemical Quenching   Chemical quenching is caused by the presence of materials in the scintillation cocktail which interfere with the transfer of energy from solvent to fluor molecules and, consequently, reduce the amount of fluorescence from the cocktail. Chemical quenching may be caused by one or more of the following processes[3]:

(1) Acid quenching resulting from interaction of H^+ with the primary or secondary fluor, both relatively strong bases. This interaction reduces the transfer of energy from solvent to fluor.
(2) Concentration quenching resulting from excessive concentration of one component (usually the primary fluor) of the cocktail. This excessive concentration interferes with energy transfer.
(3) Dilution quenching resulting from the presence of the sample or other material which does not participate in the light production process but which reduces the efficiency of energy transfer by increasing the average distance between solvent molecules.
(4) Dipole-dipole quenching resulting in nonradiative loss of energy as a consequence of an increase in the vibrational energy of some component of the cocktail. Examples of dipole-dipole quench agents include oxygen and nitromethane.
(5) Electron capture quenching resulting in interference with the transfer of energy from secondary electrons to solvent molecules. This interference is caused by capture of secondary electrons by a material with a high affinity for electrons. Halogenated hydro-

carbons such as CCl_4 and $CHCl_3$ are examples of electron capture quench agents.

Oxygen is perhaps the most common quench agent in liquid scintillation cocktails. The effects of this agent can be minimized by purging the counting vial with argon, carbon dioxide, or nitrogen, or by degassing the solution by ultrasonic agitation.

Color Quenching. A color quench agent is any colored material in the scintillation cocktail which absorbs light photons emitted by the primary or secondary fluor. Color quenching often may be reduced by bleaching or decolorizing the colored material or, if the color is not very intense, by diluting the cocktail with additional counting solution. For example, blood and samples of tissue solubilized with Hyamine 10-X have been bleached with peroxide or sodium borohydride[4].

Optical Quenching. Optical quenching describes the absorption of light photons by condensation, fingerprints, or residue on the counting vial. To reduce optical quenching, vials should be handled by the top and bottom and should be wiped clean before counting.

Quenching of a liquid scintillation sample reduces the number of light photons striking the photocathodes and, consequently, decreases the size of each voltage pulse from the photomultiplier tubes. For samples labeled with nuclides other than ^3H, this reduction in pulse height causes a shift of the pulse-height spectrum downscale towards smaller pulse heights (Figure 15-7). Because of the small number of light photons released,

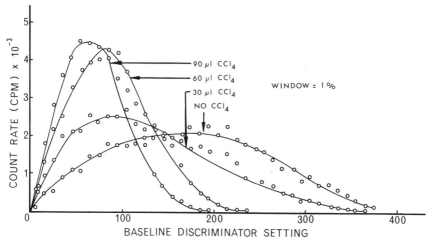

Figure 15-7 Influence of the chemical quench agent CCl_4 on the pulse-height spectrum for ^{14}C.

quenching in a scintillation cocktail containing ^{3}H reduces the number of interactions which liberate photoelectrons simultaneously in both photo-cathodes. Hence, more events are rejected by the coincidence circuit, and the height of the pulse-height spectrum is suppressed, often without a particularly noticeable shift of the spectrum downscale (Figure 15-8).

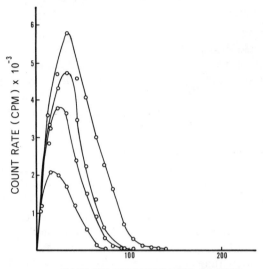

BASELINE DISCRIMINATOR SETTING

Figure 15-8   Influence of the chemical quench agent CCl$_4$ on the pulse-height spectrum for ^{3}H, illustrating the suppression of the pulse-height spectrum with increasing amounts of quench agent.

Quench Corrections

Some degree of quenching occurs in every sample prepared for liquid scintillation counting. Consequently, a procedure is needed to estimate the counting efficiency for each sample individually. Four of the more common methods are described here.

Internal Standardization.   After a sample has been counted, a very small quantity (such as a few lambdas) of a standard containing a known activity of the same nuclide is added to the sample vial. Then, the sample is re-counted and the counting efficiency is estimated as

counting efficiency

$$= \frac{\text{(count rate for sample plus standard)} - \text{(count rate for sample)}}{\text{disintegration rate for standard}}$$

Standards which are available commercially include toluene, water, and *n*-hexadecane labeled with ³H, toluene, *n*-hexadecane, and benzoic acid labeled with ¹⁴C, toluene containing elemental ³⁵S, and chlorobenzene labeled with ³⁶Cl. The counting efficiency should not be estimated by internal standardization without verifying that addition of the standard does not influence the counting efficiency[5]. Internal standardization is time consuming and subject to inaccuracies of pipetting small quantities of liquid. Also, this technique destroys the original sample.

Example 15-2
A sample provides 5243 cpm and 12231 cpm before and after a 0.00874 μCi standard is added. What is the counting efficiency for the sample?
The disintegration rate of the standard is $(0.00874 \, \mu\text{Ci})(2.2 \times 10^6 \, \text{dpm}/\mu\text{Ci})$ or 19200 dpm. The counting efficiency is

counting efficiency

$$= \frac{(\text{count rate for sample plus standard}) - (\text{count rate for sample})}{\text{disintegration rate for standard}}$$

$$= \frac{(12231 \text{ cpm}) - (5243 \text{ cpm})}{19200 \text{ dpm}}$$

$$= 0.363$$

Extrapolation to Zero Quench Agent. The count rate for a sample in the absence of quenching may be estimated by measuring the count rate for various concentrations of a quench agent and extrapolating the curve of count rate versus concentration to a concentration of 0 (Figure 15-9). For the extrapolation method of correcting for the influence of quenching, count rate measurements should be made with a series of separate samples containing varying amounts of quench agent. The addition of an increasing amount of quench agent to the same counting vial has produced erroneous estimates of counting efficiency and is not recommended[6].

Channels Ratio. The channels ratio method for estimating counting efficiency relies upon the shift of the pulse-height spectrum downscale to smaller pulse heights when a quench agent is present in the scintillation cocktail. The extent of the spectral shift and, consequently, the degree of quenching, are reflected in a change in the ratio of count rates in two counting channels. For ¹⁴C-labeled samples, one channel usually encompasses the entire pulse-height spectrum, and the second channel usually includes either the upper or lower third of the spectrum. Two adjacent nonoverlapping channels usually are used for ³H. For a particular sample, the counting efficiency is estimated by measuring the

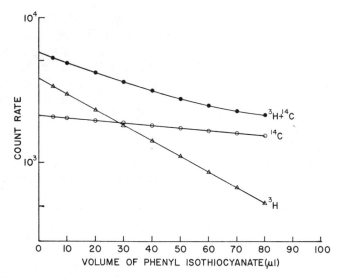

Figure 15-9 The extrapolation method for estimating the count rate for a sample in the absence of quenching. The composite ^3H–^{14}C curve has been resolved into curves for ^3H and ^{14}C. These curves illustrate the greater susceptibility of ^3H-labeled samples to changes in counting efficiency caused by quenching. Phenyl isothiocyanate is a strong chemical quench agent. [From C. Peng. 1964. Correction of quenching in liquid scintillation counting of homogeneous samples containing both carbon-14 and tritium by extrapolation method. Anal. Chem. 36:2456.]

channels ratio for the sample and then referring to a calibration curve of counting efficiency versus channels ratio.

The calibration curve is obtained by preparing a series of samples with known activity but with different amounts of a quench agent similar to that in the samples of unknown activity. Appropriate counting channels are established and the channels ratio is determined for each of the standard samples. For each standard sample, the counting efficiency is determined by dividing the count rate in one of the channels (for example, the count rate in the wide window channel for ^{14}C-labeled samples) by the disintegration rate for the sample. A plot of counting efficiency as a function of channels ratio establishes the calibration curve (Figure 15-10). Many quench agents provide similar calibration curves; nevertheless, the calibration curve obtained for one quench agent probably should not be used for another without verification that the curve is applicable[7]. Also, a calibration curve measured for one liquid scintillation counter should not be used with data from another counter.

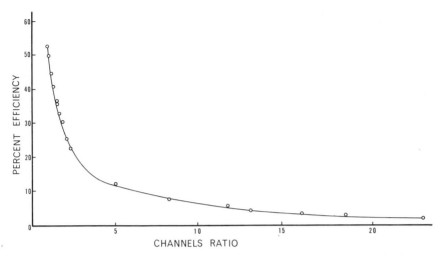

Figure 15-10 Typical calibration curve (^{14}C in toluene, quenched with CCl$_4$) for channels ratio approach to determination of counting efficiency.

External Standardization. Most of the newer liquid scintillation counters contain a γ-emitting source (such as ^{137}Cs, ^{133}Ba, or radium) which can be positioned adjacent to the sample vial in the counting chamber. During interactions of the γ-rays with the vial and the scintillation cocktail, electrons are released which produce an amount of fluorescence which varies with the degree of quenching. The fluorescence is recorded in two counting channels and the ratio of counts in the two channels is computed. This ratio, referred to as the external standards ratio (ESR), is independent of sample volume.

For a sample of unknown activity, the counting efficiency is estimated by measuring the count rate for the sample and then positioning the external standard adjacent to the sample vial. With the external standard in place, the ESR is determined and the counting efficiency is determined from a calibration curve of counting efficiency versus ESR. The calibration curve is determined by measuring the count rate and ESR for a series of samples of known activity which are quenched with an agent similar to that in the samples to be assayed. A typical calibration curve is shown in Figure 15-11.

Other methods for determining the counting efficiency for liquid scintillation samples, including a technique for estimating the counting efficiency for nonhomogeneous samples, are discussed by Peng[6]. Rogers and

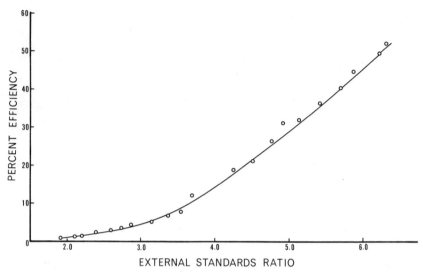

Figure 15-11   Typical calibration curve (^{14}C in toluene, quenched with CCl_4) for ESR approach to determination of counting efficiency.

Moran[8] have concluded that the careful addition of an internal standard to a scintillation cocktail is the most accurate technique for determination of counting efficiency. The channels ratio method may be as accurate as internal standardization when used with high-activity samples that are moderately quenched. These authors suggest that external standardization often is less accurate than either internal standardization or the channels ratio method.

Double Isotope Counting

More than one counting channel is available in most liquid scintillation counters, and samples containing more than one β-emitting nuclide (such as ^3H and ^{14}C) often are counted by liquid scintillation. Provided that the maximum β energies of the two nuclides are sufficiently different, the contribution of each nuclide to the total activity of the cocktail may be determined. Shown in Figure 15-12 are pulse-height spectra for ^3H and ^{14}C measured with a liquid scintillation counter with logarithmic amplification.

By selection of an appropriate window for the upper counting channel, ^{14}C can be counted with at least 50% efficiency without any contribution from ^3H in the sample. However, ^3H in the sample cannot be counted

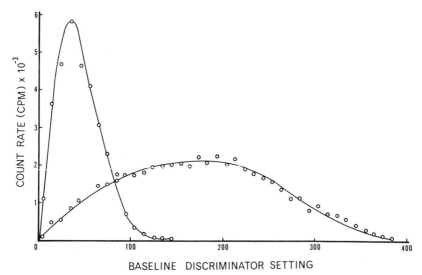

Figure 15-12 Pulse-height spectra for ^{3}H and ^{14}C in toluene measured with a liquid scintillation counter with logarithmic amplification.

without interference from ^{14}C. Consequently, the count rate for ^{3}H must be corrected for the contribution of ^{14}C to the count rate in the lower channel. In practice, windows for the two nuclides usually are selected so that the count rate in the upper channel contains a small contribution from the lower energy nuclide. For example, the data in Table 15-1

TABLE 15-1 *Counting Efficiencies for ^{3}H and ^{14}C in the Upper Channel as the Lower Discriminator of the Upper Chanel Is Moved Downscale to Include an Increasing Contribution of ^{3}H to the Count Rate in the Upper Channel*

^{3}H Counting efficiency	^{14}C Counting efficiency
0.00	50.9
0.01	57.5
0.05	62.5
0.10	65.0
0.50	70.3
1.00	72.5

[a]From Y. Kobayashi and D. Maudsley. 1970. Practical aspects of double isotope counting. *In* E. Bransome (ed.), The current status of liquid scintillation counting. Grune and Stratton, New York. p. 76.

suggest that a large increase in ¹⁴C counting efficiency can be achieved by increasing the width of the upper channel window to include a small contribution from ³H to the count rate.

A few techniques have been developed for selecting the window for the lower channel. In one method, the counting efficiency for the lower energy nuclide in the lower channel is plotted as a function of the counting efficiency for the higher energy nuclide in the same channel (Figure 15-13). The optimum window for the lower channel is the window which provides the greatest ratio of counting efficiencies for the lower and higher energy nuclides. In Figure 15-13, the maximum ratio is indicated by the point of divergence of the curve from a straight line passing through the origin. A window adjusted for this ratio provides counting efficiencies of 25% for ³H and 5% for ¹⁴C.

A second approach to selection of the window for the lower channel requires construction of an Engberg plot[9]. With this technique, the counting efficiencies for ³H and ¹⁴C are measured for a fixed window as the amplifier gain is varied from 0 to its maximum value. As shown in Figure 15-14, the counting efficiency for the higher energy nuclide is plotted on full logarithmic paper as a function of the efficiency for counting the lower energy nuclide. The optimum window for the lower channel is indicated by the point of divergence of the curve from a 45° line drawn from the base of the right side of the curve. The development

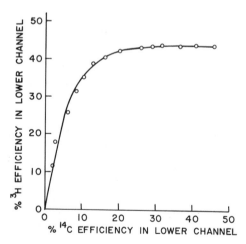

Figure 15-13 ³H Efficiency versus ¹⁴C efficiency in the lower channel for air-quenched samples in toluene–PPO. The optimum window for the lower channel is the window which provides the greatest ratio of ³H efficiency to ¹⁴C efficiency.

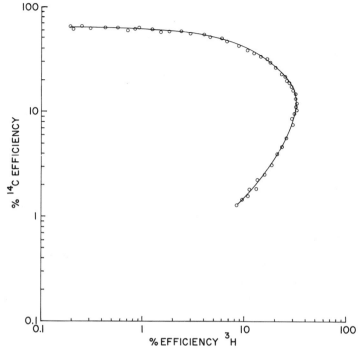

Figure 15-14   Engberg plot for air-quenched ³H and ¹⁴C samples in toluene–PPO.

of Engberg plots for other combinations of nuclides is discussed by Kobayashi and Maudsley[9].

The activity of ^{3}H and ^{14}C in a sample labeled with both nuclides may be determined with the equations

activity ^{14}C $= \dfrac{N_U - N_L(h_U/h_L)}{c_U - c_L(h_U/h_L)}$

activity ^{3}H $= \dfrac{N_L - N_U(c_L/c_U)}{h_L - h_U(c_L/c_U)}$

N_L = net total counts lower channel
N_U = net total counts upper channel
c_L = ^{14}C efficiency lower channel
c_U = ^{14}C efficiency upper channel
h_L = ^{3}H efficiency lower channel
h_U = ^{3}H efficiency upper channel

Automatic Sample Changers

Although a few single sample liquid scintillation counters are marketed commercially, most counters provide an automatic sample changer for 100 or more sample vials. One type of changer utilizes an endless belt for transport of sample vials, with each vial loaded individually into a cylinder in the belt. Each sample vial is placed into counting position when its cylinder moves into position over the elevator shaft. A second type of changer provides sample trays which are placed into shelves in the counter console. An elevator positions the trays for sample exchange and counting. This changer may be programmed to select trays in any sequence desired, and to provide instrument settings appropriate for the samples in a particular tray.

REFERENCES

[1] Rapkin, E. 1970. Development of the modern liquid scintillation counter. *In* E. Bransome, Jr. (ed.), The current status of liquid scintillation counting. Grune and Stratton, New York. p. 45.

[2] Ross, H. 1967. The balanced quenching method for counting carbon-14, Internat. J. Appl. Radiat. & Isotopes 18:335.

[3] Neary, M., and A. Budd. 1970. Color and chemical quench. *In* E. Bransome, Jr. (ed.), The current status of liquid scintillation counting. Grune and Stratton, New York. p. 273.

[4] Rapkin, E. 1967. Preparation of samples for liquid scintillation counting. *In* G. Hine (ed.), Instrumentation in nuclear medicine. Academic Press, New York. p. 181.

[5] Hendee, W., G. Ibbott, and K. Crusha. 1972. ^{3}H-toluene, ^{3}H-water, and ^{3}H-hexadecane as internal standards for toluene- and dioxane-based liquid scintillation cocktails. Internat. J. Appl. Radiat. Isotopes 23:90.

[6] Peng, C. 1970. A review of methods of quench correction in liquid scintillation counting. *In* E. Bransome, Jr. (ed.), The current status of liquid scintillation counting. Grune and Stratton, New York. p. 283.

[7] Bush, E. 1963. General applicability of the channels ratio method of measuring liquid scintillation counting efficiencies. Anal. Chem. 35:1024.

[8] Rogers, A., and J. Moran. 1966. Evaluation of quench correction in liquid scintillation counting by internal, automatic external, and channels' ratio standardization methods. Anal. Biochem. 16:206.

[9] Kobayashi, Y., and D. Maudsley. 1970. Practical aspects of double isotope counting. *In* E. Bransome, Jr. (ed.), The current status of liquid scintillation counting. Grune and Stratton, New York. p. 76.

16

Autoradiography and Radiochromatography

Autoradiography

A photographic emulsion consists of grains of silver halide (usually silver bromide) suspended in a gelatin matrix. Almost always, the emulsion is coated upon a glass plate or upon a film base of cellulose acetate or polyester resin. When the emulsion is exposed to light or ionizing radiation, electrons are released which initiate the reduction of silver ions to metallic silver at the locations where the radiation interacts in the emulsion. When the emulsion is placed in developer, the metallic silver atoms catalyze the reduction of additional silver ions to metallic silver in the immediate vicinity of the interaction sites. By this process, metallic silver is deposited on the glass plate or film base in a pattern which reflects the distribution of radiation interactions in the emulsion. Silver halide granules which are undisturbed by the incident radiation are removed by a fixing solution containing sodium thiosulfate or ammonium thiosulfate. The optical density or "blackening" of any region of the processed film depends upon the amount of metallic silver present and reflects the energy deposited in the region by the incident radiation.

Autoradiography refers to the use of photographic methods to determine the distribution of radioactivity in a specimen containing radioactive material. The specimen is placed in contact with a photographic emulsion, and, after a suitable exposure, the emulsion is processed. The distribution of metallic silver in the resulting autoradiograph indicates the distribution

217

of radioactivity in the specimen. Various methods have been established to ensure intimate contact between emulsion and specimen during exposure of the emulsion to radiation.

Temporary Contact Autoradiography

With this method, an emulsion mounted on a film base is placed in firm contact with the specimen for a predetermined exposure time. After exposure, the specimen and film are separated and the film is processed to reveal the distribution of radioactivity in the specimen. Usually, x-ray film (nonscreen or dental x-ray film) is used. The temporary contact method is particularly useful for gross anatomic sections, histologic sections mounted on glass slides, and paper and thin-layer chromatograms. Compared to permanent contact techniques, temporary contact autoradiography furnishes poor resolution, because intimate contact is difficult to achieve between the emulsion and all portions of the specimen.

For specimens labeled with ^{3}H or ^{14}C, the exposure time may be several weeks for an autoradiograph prepared by the temporary contact technique. The exposure time sometimes can be reduced by spraying the specimen with a scintillation solution prior to contact with the x-ray film. Alignment of the original specimen with the processed autoradiograph sometimes is difficult, particularly when the specimen has shrunk during contact with the emulsion. Because of alignment difficulties and inferior resolution, temporary contact autoradiographs are useful only when high resolution is not required.

Permanent Contact Autoradiography

To improve resolution and reduce problems of alignment, histologic sections containing radioactive material may be mounted permanently in contact with a photographic emulsion. Usually, nuclear emulsions such as those in Eastman Kodak's NTA or NTB series are used for permanent contact autoradiography. Compared to emulsions on x-ray film, nuclear emulsions have a greater concentration of silver bromide granules of small and very uniform size. Consequently, these emulsions provide high resolution and a reasonable sensitivity to β particles. After exposure, the emulsion is developed to precipitate metallic silver directly on the section, and the deposited silver is viewed simultaneously with the section by either bright-field or phase-contrast microscopy. The distribution of radioactivity in the specimen may be estimated quantitatively by counting silver grains in various regions of the specimen. A variety of techniques have been developed for permanent contact autoradiography.

Emulsion Coating. Fine grain emulsions which have been melted by heating may be poured or painted over a histologic specimen mounted on a glass slide. Alternately, the specimen may be dipped into a vessel containing liquid emulsion. After the emulsion has hardened, the specimen and emulsion are placed in developing and fixing solutions. The resolution of the resulting autoradiograph is limited primarily by the variable thickness of the coated emulsion. More uniform thicknesses of emulsion are obtained by dipping the specimen into the emulsion. With the emulsion coating technique, care must be taken to prevent shrinkage of the specimen or emulsion and the presence of bubbles in the emulsion.

Specimen Flotation. Histologic sections floating on water may be captured on a glass slide coated with emulsion which is brought up underneath the section. After an appropriate exposure time, the specimen is stained and the emulsion is developed and fixed. Photographic solutions may not penetrate well through some parts of the specimen, and development may not be uniform; also, the photographic solutions may damage the specimen. Reasonable resolution ($5-7\,\mu$) may be achieved by the technique of specimen flotation, provided care is taken to prevent the formation of wrinkles or other distortions in the specimen. Although the specimen flotation procedure must be conducted in the dark, the procedure is simple and can be employed with a wide variety of emulsions.

Emulsion Flotation. Thin ($4-5\,\mu$) layers of nuclear emulsion stripped from a special base may be floated on water and captured by a stained specimen mounted on a glass slide. Compared to techniques discussed previously, this method provides improved resolution ($1-3\,\mu$) and usually is preferred when intracellular localization or other autoradiographic studies requiring high resolution are desired. To protect the specimen from photographic solutions, the specimen may be coated with a thin layer ($\sim 1\,\mu$) of 1% collodion in equal parts of ether and alcohol prior to capture of the emulsion. One disadvantage of the emulsion flotation technique is the long exposure time required because of the low sensitivity of stripping films.

Electron Microscopic Autoradiography

Relatively recently, techniques have been developed for electron microscopic autoradiography at very high levels of magnification. An ultrathin specimen embedded in plastic is coated with a very fine grain emulsion (grain diameters of $0.03-0.1\,\mu$) by dipping a fine copper grid containing the specimen into liquid emulsion. Alternately, a metal loop may be dipped into liquid emulsion and the captured emulsion deposited over the grid containing the specimen. A resolution of $0.1-0.3\,\mu$ is obtainable by electron microscopic autoradiography, sometimes referred to as "molecular autoradiography." Procedures for electron microscopic autoradiography are described in the literature[1–3].

Autoradiographic Resolution and Artifacts

The resolution furnished by an autoradiographic technique may be described as the minimum distance required between two point sources of radioactivity for the sources to be resolved as two distinct images in the autoradiograph. In general, the resolution is improved as the following factors are reduced: emulsion thickness and grain size, specimen thickness, distance between specimen and emulsion, and range of the radiation.

The estimated mean ranges of β particles in a photographic emulsion are listed in Table 16-1 for a few nuclides of biologic interest. Tritium furnishes autoradiographs of superior resolution; higher energy β particles penetrate the emulsion to greater depths and provide images which are less distinct and less useful for studies such as intracellular localization. Similarly, x- and γ-rays usually provide autoradiographic images of inferior quality, although soft x-rays such as those from ^{55}Fe can produce images of acceptable quality when used with special emulsions. Because higher specific activities can be obtained by labeling compounds with ^{3}H rather than ^{14}C, and because ^{3}H provides superior resolution, this nuclide is preferred for most autoradiographic studies.

Autoradiographic images are susceptible to a variety of artifacts related to inappropriate care of the specimen or emulsion. For example, the specimen must be stained and dehydrated carefully to prevent the radioactive material from leaching from the specimen or changing its distribution within the specimen. For specimens coated with an emulsion prior to staining, the staining process should not include materials which are absorbed by the emulsion. For example, hematoxylin–eosin stain is absorbed strongly by many nuclear emulsions. Artifacts in the autoradiographic image may be caused also by conditions such as volatile agents

TABLE 16-1 *Estimated Mean Range of β Particles in a Photographic Emulsion*

Nuclide	Average β energy, MeV	Range, μ
^{3}H	0.0055	1
^{14}C and ^{35}S	0.049	10
^{131}I	0.188	100
^{32}P	0.70	800

in the specimen, mechanical pressure on the emulsion, the presence of light, dust and debris during exposure and processing of the autoradiograph, and expansion or contraction of the specimen or emulsion during exposure and processing. The presence of some artifacts may be revealed by preparing duplicate "autoradiographs" for specimens which contain no radioactivity.

Exposure Time

Exposure times in autoradiography are largely empirical, and usually are determined by exposing a series of samples for varying intervals of time. To obtain an autoradiograph of satisfactory density with nonscreen x-ray film, an exposure time should be selected which enables 10^6–10^8 β particles to strike each square centimeter of emulsion. An estimate of exposure time for another type of emulsion should include correction for the variation in sensitivity between the emulsion and nonscreen x-ray film.

The exposure time for an autoradiograph should not be too long, because chemical and background fogging of the image increases with exposure time. Also, the latent image (radiation-induced changes in the emulsion which cause the deposition of metallic silver when the emulsion is developed) tends to fade with time. The rate of fading of the latent image varies from one emulsion to another and is described by the half-life of the latent image. This half-life is defined as the time required for the optical density of an autoradiograph to decrease to half because of latent image fading. The latent image half-life is many months for ordinary x-ray film, about 1 month for most nuclear emulsions, and only 10 days or so for some emulsions. For most autoradiographic techniques, the exposure time should be no greater than twice the half-life of the radioactive material or the latent image, whichever is shorter.

Radiochromatography

Chromatography is a technique to separate the components of a mixture of compounds. The chromatographic procedure chosen to separate compounds of a particular mixture depends upon the physical and chemical properties of the compounds. In paper chromatography (PC), for example, compounds are separated because they migrate in paper at different velocities. In this technique, a few drops of a solution of the mixture of compounds are deposited and dried near one end of a suitable paper. This end of the paper is placed in a selected solvent or mixture of solvents. The solvent migrates in the paper by capillary action and transports the compounds at different velocities through the paper. By placing an adjacent side of the paper in a different solvent, additional separation of the compounds may be achieved. The final location of each compound in the paper chromatogram may be detected by the color, a chemical reaction, ultraviolet absorption, or fluorescence. If the compounds are radioactive, their final locations may be determined with a suitable radiation detector.

Thin-layer chromatography (TLC) resembles paper chromatography (PC), except that glass plates containing a thin layer of absorbent porous material (such as silica gel, powdered cellulose, or aluminum oxide) are used in place of paper. In general, paper chromatography is used for hydrophilic compounds (for example, amino acids, sugars and purine derivatives), and thin layer chromatography is employed for separation of lipophilic compounds such as fatty acids and steroids[4]. Both techniques provide chromatograms with the compounds of the mixture separated spatially.

If the compounds to be separated are volatile, then they can be volatized and transported in a stream of inert gas through a column of absorbent material. The compounds are separated by partition and absorption during transport, and reach the outlet of the column at different times. Consequently, this procedure, referred to as gas chromatography (GC), provides a temporal distribution rather than a spatial distribution of compounds. At the column outlet, the separated compounds may be analyzed by hydrogen flame ionization, electron capture, argon ionization, thermal conductivity, density balance, or, if radioactive, by a suitable radiation detector.

In liquid chromatography (LC), a solution of the compounds to be separated is dropped onto the upper end of a column composed of an absorbent material such as aluminum oxide, silica gel, ionic resins, or powdered sugars. By adding a solvent or mixture of solvents to the upper end of the column, the compounds are transported at different rates through the column and reach the outlet of the column at different times.

Proteins and other compounds which can be ionized by a technique such as dissolution in a solvent with an appropriate pH may be separated by application of an electric field. This technique is referred to as *electrophoresis*.

The detection of compounds separated chromatographically is made easier if the compounds are labeled with a radioactive nuclide. The nuclides used most often are ^3H and ^{14}C. Because of the increased mass of ^3H compared to ordinary hydrogen, tritiated compounds may be transported more slowly than their nonradioactive counterparts. This possible difference in transport rate should be considered during interpretation of results.

Paper and Thin-Layer Radiochromatography

Labeled compounds separated by PC or TLC usually are localized by recording the count rate as the chromatogram is moved slowly past a radiation detector. For certain applications, this technique does not furnish satisfactory results; in these cases, a destructive method must be employed for localization of radioactivity on the chromatogram. For example, the counting efficiency usually is less than 1% for external scanning of a chromatogram containing ³H-labeled compounds. The efficiency for detection of tritium may be increased 10–20 times by cutting the chromatogram into sections and counting each section in a liquid scintillation counter.

A typical external scanner for paper radiochromatograms accepts paper strips up to 2 in. wide and moves the strips between two windowless gas flow GM tubes or proportional detectors. The use of two detectors not only improves the counting geometry, but also compensates for variations in absorption of the radiation in the sample or the paper. A windowless gas flow detector is essential for chromatograms containing ³H, because even the thinnest window absorbs the low-energy β particles from this nuclide.

The count rate furnished by the detector is displayed on a ratemeter

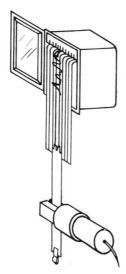

Figure 16-1 Simple procedure for external scanning of paper radiochromatograms. The chromatogram is attached to the paper of the strip chart recorder and both move at constant speed, permitting easy alignment of the chromatogram and the recording of count rate.

and recorded on a strip chart recorder which moves at the same rate as the paper chromatogram. The spots of activity on the chromatogram are recorded with a resolution which improves with decreasing width of the slit in front of each detector. Typical slit widths are 2.5, 5, 7.5, and 10 mm. The width of the slits always should be smaller than the width of the radioactive spots on the chromatogram. The length of the slits may vary from 10 to 40 mm in a direction perpendicular to the motion of the chromatogram. The chromatogram passes between the slits at a variable speed (for example, 1–200 cm/hr). A simple chromatogram scanner is illustrated in Figure 16-1; more sophisticated scanners are available commercially which provide automatic data recording and which can accommodate a continuous chain of paper strips up to 50 m in length. A scan of a chromatogram for ^{3}H-thymidine is shown in Figure 16-2. The area under each peak is proportional to the activity included within the corresponding spot in the chromatogram.

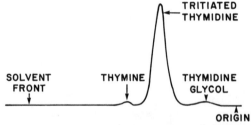

Figure 16-2 Scan of a paper chromatogram for tritiated thymidine which reveals the presence of the contaminants thymine and thymidine glycol. The chromatogram was obtained with ethyl acetate, formic acid, and water solvents (60:5:35 v/v/v), a scan speed of 1 cm/min, and a slit width of 2.5 mm. [From W. Briner. 1966. Quality Control, pyrogen testing, and sterilization of radioactive pharmaceuticals. *In* G. Andrews, R. Kniseley, and H. Wagner (ed.), Radioactive pharmaceuticals CONF-651111. Clearinghouse for Federal Scientific and Technical Information, Springfield, Va. p. 93.]

Low-energy β particles cannot penetrate the glass plate of a thin-layer radiochromatogram, and only one radiation detector is used to scan these chromatograms. The distance between the detector and the glass plate usually is no greater than 0.5 mm. For certain applications, segments of the absorbent layer of a thin-layer chromatogram must be scraped from the glass plate and counted as a suspension in a liquid scintillation counter. Some compounds can be eluted from the scrapings and counted in solution rather than in suspension.

Some PC scanners can be modified to accept thin-layer chromatograms, provided that the width of the TLC plates is not greater than 2 in. For wider TLC plates, a two-dimensional external scanner must be used. This instrument is useful also for scanning two-dimensional paper chromatograms and one-dimensional paper chromatograms wider than 2 in. With most two-dimensional scanners, the limits of longitudinal motion, the distance between adjacent scan lines, and the scanning speed can be varied over wide ranges.

The counting efficiency E for external scanning of a chromatogram is the ratio of the count R in cpm to the disintegration rate A in dpm for a sample centered in the detector slit. The total number N of counts furnished by a radioactive spot is

$$N = EA\frac{d}{v}$$

where d is the slit width in mm and v is the scan speed in millimeters per minute.

Example 16-1
What is the activity of a radioactive spot providing 1120 counts when scanned with a slit width of 5 mm and a scan speed of 10 mm/min, if the counting efficiency is 1.5%?

$$N = EA\frac{d}{v}$$

$$A = \frac{Nv}{Ed}$$

$$= \frac{(1120)(10 \text{ mm/min})}{(0.015)(5 \text{ mm})}$$

$$= 149,000 \text{ dpm}$$

Since $1 \mu Ci = 2.2 \times 10^6 \text{ dpm}$

$$A = \frac{149,000 \text{ dpm}}{2.2 \times 10^6 \text{ dpm}/\mu Ci}$$

$$= 0.068 \mu Ci$$

For samples emitting low-energy β particles, the counting efficiency decreases rapidly with increasing thickness of the radioactive spot on a paper or thin-layer chromatogram. For a sample thickness of 10 mg/cm^2 and a single windowless gas-flow detector, for example, counting efficiencies of about 15% for ^{14}C and 0.7% for ^{3}H can be anticipated[4].

These efficiencies have been improved by impregnating the chromatogram with a scintillator and measuring the light from the scintillator with a photomultiplier tube[5].

Gas and Liquid Radiochromatography

Radiochromatograms obtained with paper or thin-layer plates provide a spatial distribution of radioactivity and may be counted at any convenient time after completion of the chromatographic separation. In gas and liquid chromatography, the radioactivity is distributed in time and usually is monitored by a radiation detector as the gas or liquid stream leaves the chromatographic column. Alternately, individual fractions of the effluent may be collected for analysis at a later time. Although collection of individual fractions is tedious, fraction collectors still are used occasionally, particularly for some liquid chromatographic procedures where radiation detectors for continuous monitoring have several disadvantages.

The total number N of counts obtained as a bolus of activity traverses the sensitive volume of a detector is

$$N = EA \frac{V}{f}$$

where E and V are the efficiency and sensitive volume of the detector, A is the activity of the bolus, and f is the flow rate of the gas or liquid stream in cubic centimeters per minute. Counting efficiencies as high as 70% for ^3H and 85% for ^{14}C have been observed during continuous monitoring of the effluent from a gas chromatograph.

Example 16-2

What is the ^3H activity in a bolus of radioactive gas from a gas chromatograph equipped with a detector which furnishes a detection efficiency of 65% for ^3H? The sensitive volume of the detector is 10 cm³, the flow rate is 40 cm³/min, and the number of counts obtained is 850.

$$N = EA \frac{V}{f}$$

$$A = \frac{Nf}{EV}$$

$$= \frac{(850)(40 \text{ cm}^3/\text{min})}{(0.65)(10 \text{ cm}^3)}$$

$$= 5230 \text{ dpm}$$

$$= 0.0024 \ \mu\text{Ci}$$

Occasionally, the effluent from a gas chromatograph may be mixed with counting gas and counted directly in a gas-flow GM or proportional counter. More often, the effluent produces spurious counts or severe quenching in the detector, and must be transformed chemically before mixing with the counting gas. Two procedures, an oxidative method and hydrogenation cracking, are employed routinely for the chemical transformation[4].

The oxidative method is useful for ^{14}C-labeled compounds containing carbon, hydrogen, and oxygen, but is unreliable if other elements such as nitrogen, halogens, or sulfur are present. "Memory effects" are observed occasionally when ^{3}H-labeled substances are transformed by the oxidative method. The method of hydrogenation cracking is more reliable, and has been used with a wide variety of substances[6]. If destruction of the sample by chemical transformation is undesirable, then a flow-through ionization chamber may be employed as a radiation detector. In some cases, these detectors also require chemical transformation of the sample.

Scintillation flow detectors described in Chapter 14 also are used to monitor the effluent from gas chromatographs. Flow detectors are the only detectors routinely suitable for on-line monitoring of samples obtained by liquid chromatography. Poor counting efficiencies, problems with effluents which react with the scintillator, and quenching caused by colored solutions reduce the usefulness of scintillation flow detectors for some applications. For these applications, an automatic fraction collector is used.

PROBLEMS

1. What is the activity of a radioactive spot providing 940 counts when scanned with a slit width of 2.5 mm and a scan speed of 3 mm/min? The counting efficiency is 1%.

2. What is the activity in a bolus of tritium from a gas chromatograph? The detector provides 1025 counts and has a sensitive volume of 20 cm^3. The efficiency of the detector is 50% for tritium, and the flow rate of the gaseous effluent is 20 cm^3/min.

REFERENCES

[1] Caro, L. 1964. High-resolution autoradiography. *In* D. Prescott (ed.), Methods in cell physiology, vol. 1. Academic Press, New York. p. 327.

[2] Revel, J., and E. Hay. 1963. An autoradiographic and electron microscopic study of collagen synthesis in differentiating cartilage. Z. Zellforsch Mikroskop Anta. 61:110.

[3] Salpeter, M., and L. Bachmann. 1964. Autoradiography with the electron microscope. A procedure for improving resolution, sensitivity, and contrast. J. Cell Biol. 22:469.

[4] Berthold, F., and M. Wenzel. 1967. Radiochromatographic counting techniques. *In* G. Hine (ed.), Instrumentation in nuclear medicine. Academic Press, New York. p. 251.

[5] Roucayrol, J., J. Bergner, G. Meyniel, and J. Perrin. 1964. Etude comparée de la mesure des activités bêta sur couches chromatographiques minces: influence de détecteur, de la nature de la couche mine et de l'energie du rayonnement beta. Internat. J. Appl. Radiat. Isotopes 15:671.

[6] Simon, H., G. Muellhofer, and R. Medina. 1965. A general method for the rapid determination of carbon-14- and hydrogen-3-labeled substances by gas chromatography. Proc. Symp. Radioisotope Sample Measurement Tech. Med. Biol., IAEA, Vienna. p. 317.

17
Activation
Analysis

By bombardment with neutrons, charged particles, and high-energy photons, stable atoms can be transformed into radioactive isotopes. By measuring the radiation emitted by a bombarded sample, the presence of one or more elements of particular interest in the sample often can be determined quantitatively. This approach to detection of trace elements and analysis of the chemical composition of a sample is referred to as *activation analysis*.

Radiation Sources for Activation Analysis

Activation analysis of a sample may be performed with any type of radiation which transforms stable atoms into radioactive isotopes. For example, slow and fast neutrons, protons, deuterons,* tritons,† α particles, and high-energy x and γ-rays have been employed for activation analysis. Most often, samples are activated with neutrons, especially slow (thermal) neutrons.

Most elements with Z above about 10 exhibit reasonable cross sections for activation by thermal neutrons. Usually, the activation is achieved by an (n,γ) reaction, and the product nucleus is a radioactive isotope of the target element. The four major components (H, C, N, and O) of biologic tissue have atomic numbers less than 10 and are activated

---

*Deuteron = nucleus of deuterium.
†Triton = nucleus of tritium.

only rarely by thermal neutrons. Consequently, thermal neutrons are useful for analysis of trace elements in biologic tissue, because the higher Z trace elements can be activated and detected without interference from activated products of the four low-Z elements which constitute the major portion (96%) of biologic tissue. The availability of high-flux densities (up to 10^{15} neutrons/cm²-sec) of thermal neutrons in nuclear reactors also contributes to the widespread use of these particles for activation analysis.

A few elements, especially those with Z less than about 10, offer no suitable reactions for activation by thermal neutrons. These elements can be activated with fast neutrons; however, the extent of activation of low-Z elements by fast neutrons is small compared to activation of higher Z elements by thermal neutrons. The reduced sensitivity is a reflection of the small activation cross section of most elements for fast neutrons, as well as the limited flux densities of fast neutrons available from neutron generators. Because of their greater kinetic energy, fast neutrons may interact with sample nuclei in a variety of ways, and a particular radioactive isotope may be produced by transformation of stable atoms of more than one element in the sample. Consequently, the determination of sample composition often is more complicated when the sample is activated by fast neutrons.

High-energy charged particles, particularly deuterons, also have been used occasionally for activation of biologic samples. The cross section for activation by charged particles is relatively low for most elements, particularly those of higher Z, and interfering reactions occur frequently. Other disadvantages of charged-particle activation include difficulties in obtaining adequate and uniform flux densities of particles, and also heating of the sample as the particles are absorbed.

Activation of samples with high-energy photons has not been investigated widely, primarily because the activation cross sections are low for most elements and because suitable sources for high-intensity beams of high-energy photons are not readily available. Wahl, Nass, and Kramer [1] have used a 15 Ci source of ^{24}Na to detect as little as 50 mg of deuterium by photoneutron (γ,n) activation of samples with the 2.75 MeV γ-rays from this isotope.

Preparation of Samples

After a sample has been irradiated, there is little chance of sample contamination by improper selection of reagents or containers. Before exposure to radiation, however, the sample must be handled with extreme care to prevent the addition of trace elements which might be activated during irradiation. As rapidly as possible, and with no more preparation than absolutely necessary, a sample to be activated should be placed in a sealed container and stored until irradiation. Plastic containers always are preferable to glass for storage of a sample before irradiation. For dissection of biologic tissues, silica knives (for example, a splinter from a broken flask) should be employed. Contact of a sample with metal always should be avoided. Although stainless steel utensils such as knives and scissors might be acceptable, most utensils are plated with chromium, zinc, or nickel which can add trace impurities to the sample. Utensils should be sterilized chemically (for example, with ethylene oxide) rather than by autoclaving, because trace elements from water and the autoclave can be transferred to the utensils. Samples should be dried in a closed system by either vacuum or infrared lamps. A drying oven should not be used, because contaminants may be vaporized onto the sample from the heating elements and the insulation of the oven.

If the temperature of a sample and its container remains below 60°C during irradiation, the sample may be sealed in polyethylene bags during activation. Aluminum canisters and aluminum foil also are acceptable sample containers, provided that no volatile substances are present in the sample. Although expensive and difficult to seal, silica ampules and tubes are insensitive to temperature and are used often as sample

TABLE 17-1 *Sources of Contamination of Biologic Samples in Activation Analysis*[a]

Observed contaminant	Suspected source	Function	Acceptable substitute
I, Hg, Mn, As	Antiseptic	Cleaning	Ethyl alcohol
Ni, Cr, Zn, Mn, Cu, Cd, Ag	Metal needles with hubs	Sample dissection and preparation	Platinum or stainless steel needles without hubs
Ni, Cr, Zn, Mn, Cu, Fe	Metal syringes	Sample dissection and preparation	Plastic syringes
Zn, Ni, Cr	Metal-plated knives and scissors	Sample dissection and preparation	Plastic, glass, stainless steel knives, and scissors
Many	Metal containers	Sample storage	Plastic containers
Pb, Ag, Cd, As, Cu, Mn	Autoclave	Sterilization	Ethylene oxide
Many	Drying ovens	Sample drying	Infrared lamp
Many	Reagents	Sample preparation	None

[a]From H. Kramer, and W. Wahl. 1968. Activation analysis. *In* H. Wagner (ed.), Principles of nuclear medicine. (Philadelphia: W. B. Saunders Co., Philadelphia. p. 811.

containers. Containers for samples activated by charged particles must be designed with attention to the low penetration of the particles.

Listed in Table 17-1 are a few contaminants and their suspected sources which have been observed during activation of biologic samples. Included in Table 17-1 are acceptable substitutes for the procedures suspected as the sources of contamination.

Sensitivity of Activation Analysis

The sensitivity of activation analysis for determination of sample composition depends upon the activity of the isotopes of interest in the sample and upon the efficiency with which the radiation emitted by the isotopes can be detected. The induced activity depends upon the neutron flux density, the cross sections for activation of the target atoms, the irradiation time, and the number of target atoms in the sample. The activity at the time of measurement of the radiation depends also upon the half-lives of the isotopes of interest and the time elapsed between activation and measurement. Often, the irradiation time and the time between irradiation and measurement are selected to permit analysis for a particular isotope with minimum interference from other radioactive isotopes in the sample. For example, if iodine in the thyroid gland is to be estimated by bombardment of the gland with thermal neutrons according to the reaction

$$^{127}\text{I}(n,\gamma)^{128}\text{I}$$

then a short irradiation time of 30–60 min should be chosen, because a reasonable amount of short-lived ^{128}I ($T_{1/2} = 25$ min) can be obtained in this period without production of much ^{24}Na ($T_{1/2} = 15$ hr). However, if the amount of sodium in biologic tissue is to be determined by the reaction

$$^{23}\text{Na}(n,\gamma)^{24}\text{Na}$$

then a longer irradiation time of perhaps 30–60 hr should be chosen. After irradiation and before estimation of the ^{24}Na activity, the sample might be stored for several hours to permit shorter lived nuclides to decay.

The sensitivity of activation analysis for quantitative measurement of various elements in a sample depends upon the facilities available for irradiation of the sample and upon the techniques employed for detection of radiation from the activated sample. Consequently, sensitivities such as those in Tables 17-2 and 17-3 should be used only as general guidelines. Sensitivities are difficult to estimate for activation by charged particles, because the flux densities and irradiation conditions available for charged-particle activation are so variable. With careful selection of the

Table 17-2 *Estimated Sensitivities for Activation of Elements by Thermal Neutrons at a Flux Density of 10^{12} n/(cm²-sec)[a]*

Sensitivity, μg	Elements
10^{-4}	Ag, As, In, Ir, Mn
10^{-3}	Al, Br, Cu, Ga, I, Na, Sb, Sc, V, W
10^{-2}	Au, Cl, Co, Cs, Ge, Hg, K, Rb, Sr, Ta, U, Y
10^{-1}	Ba, Cd, Cr, Mo, Ni, P, Se, Si, Te, Ti, Zn
1	Bi, Ca, F, Mg, Tl, Sn

[a]From J. Lenihan. 1967. Nuclear activation analysis. *In* G. Hine (ed.), Instrumentation in nuclear medicine. Academic Press, New York. p. 309.

TABLE 17-3 *Estimated Sensitivities for Activation of Elements by Fast Neutrons at a Flux Density of 10^9 n/cm²-sec; Irradiation Time 1 hr or 4 Half-Lives, Whichever is Shorter[a]*

Sensitivity, μg	Elements
1	Ag, Br, Cu, F, K, P, Sb, Si
10	Al, Cr, Mn, Mo, N, O, V, Zn
100	As, Co, I, Mg, S

[a]From J. Lenihan. 1967. Nuclear activation analysis. *In* G. Hine (ed.), Instrumentation in nuclear medicine. Academic Press, New York. p. 309.

type and energy of the bombarding particles, sensitivities in the submicrogram region often can be attained by charged-particle activation.

Analysis of Activated Samples

To permit the measurement of radiation from one isotope without interference by radiation from other isotopes, samples such as biologic tissue which contain a variety of activation products often must be manipulated chemically after irradiation. Chemical manipulation usually includes the following steps: (1) addition of a carrier; (2) digestion or other preliminary treatment; and (3) separation of the isotope of interest.

The addition of carrier to the sample permits determination of the percentage of yield of succeeding chemical processes. Also, by adding a surplus of atoms which are similar chemically to the atoms of interest in the sample, the loss of the atoms of interest by absorption is minimized. The carrier should be added at the earliest convenient moment after activation of the sample. Usually, no more than a few milligrams of carrier are added. If β particles are to be detected, then the amount of carrier should be limited to prevent excessive self-absorption of the radiation in the final mixture; if the procedure for sample preparation involves an ion exchange process, then the amount of carrier added may be limited by the saturation characteristics of the ion-exchange resin.

Chemical bonds and valence states may have changed during irradiation of the sample, and all atoms of the same element in the sample may not exhibit identical chemical behavior. To ensure that succeeding chemical procedures are effective, all irradiated and carrier atoms of interest should be placed in the same chemical form. Consequently, the sample should be subjected to a suitable oxidation or reduction process soon after addition of carrier.

After oxidation or reduction, the atoms of interest are separated from the sample by a conventional method such as volatilization, precipitation, chromatography, solvent extraction, electrodeposition, or amalgam exchange. The choice of a method for separation is influenced not only by chemical concerns, but also by consideration of the atoms in the sample which interfere most with analysis for the isotope of interest, and by the need for haste when a short-lived isotope is of interest.

After the separation of the component of interest from the sample, the radiation is detected by one of the counting techniques described earlier. If different isotopes of the same element are present in the isolated material, then radiation from the isotope of interest must be distinguished from radiation emitted by the other isotopes. This distinction often can be achieved by obtaining the composite pulse-height spectrum for γ-rays from the sample, and then stripping pulse-height spectra from the composite spectrum until only the spectrum for the isotope of interest remains. The spectral stripping process can be performed either manually or by computer.

Applications to Biology

Numerous applications of activation analysis to biologic problems are discussed in the literature[2–5]; consequently, only a few examples of the usefulness of this method are presented here.

(1) Trace elements in biologic specimens. Elements such as Zn, Cu, Mn, Mo, and Co are not major components of biologic tissue but

are important constituents of hormones and catalysts for enzyme systems. Other trace elements (such as Cd, Cr, F, Ni, Rb, Si, and V) also are present in body tissues, although their importance has not been clearly established. Analysis for these elements, which may be present in tissue at levels below the detection limit for conventional analytical techniques, is aided greatly by activation analysis. For example, activation by thermal neutrons has been used to study zinc metabolism in the body and its relation to ureteric calculus, alcoholic cirrhosis of the liver, and hypogonadism. This technique has been useful also for study of the importance of Mn, Mo, V, and other elements to the mineralization of tooth enamel and to the development of dental caries[6].

(2) Arsenic in hair. The measurement of arsenic in human hair by routine analytical procedures requires a sample of hair of about 1 g. If the arsenic is measured by activation analysis with thermal neutrons, then only a 1–2 mm length of one hair is necessary. The ^{76}As produced by (n,γ) activation of stable ^{75}As is a γ-emitting isotope with a convenient half-life of 26 hr. An interesting application of this technique is the estimation of the arsenic content in samples of hair from Emperor Napoleon I[7]. The study revealed an excessive quantity of arsenic in the hair, suggesting the possible ingestion of medical preparations or other products containing arsenic.

(3) Iodine uptake by the thyroid gland. Measurement of the uptake of radioactive iodine (^{123}I, ^{125}I, ^{131}I, ^{132}I) by the thyroid gland furnishes data about the rate of uptake and turnover of iodine in the gland (Chapter 20). However, this technique does not reveal the amount of stable iodine with which the radioactive iodine mixes. To estimate the amount of stable iodine in the body (the "iodine pool"), a sample of plasma or urine may be passed down an ion exchange column, and the stable ^{127}I captured by the column may be activated to ^{128}I by bombardment with thermal neutrons. From measurement of the ^{128}I activity on the column, the amount of stable iodine in the body can be estimated.

(4) Neutron activation analysis *in vivo*. The amounts of calcium, sodium, and chlorine in the body have been estimated by whole-body exposure of animals and humans to fast neutrons of various energies[8, 9]. Estimates of the calcium content in simulated human phantoms irradiated with 14 MeV neutrons were accurate to within 1–2%, and provided an average whole-body radiation dose of about 0.65 rem. A pulse-height spectrum for γ-rays from a patient bombarded by 14 MeV neutrons is shown in Figure 17-1.

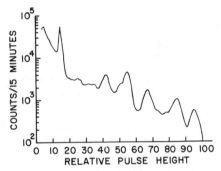

Figure 17-1 Pulse-height spectrum for γ-rays from a patient obtained 6 min after exposure to 14 MeV neutrons at a flux density of 9×10^4 n/cm²-sec. [From S. Cohn, C. Dombrowski, and R. Fairchild. 1970. *In vivo* neutron activation analysis of calcium in man. Internat. J. Appl. Radiat. Isotopes 21:127.]

REFERENCES

[1] Wahl, W., H. Nass, and H. Kramer. 1966. Use of stable isotopes and activation analysis for *in vivo* diagnostic studies. *In* G. Andrews, R. Kniseley, and H. Wagner (ed.), Radioactive pharmaceuticals CONF-651111. Clearinghouse for Federal Scientific and Technical Information, Springfield, Va. p. 191.

[2] Comar, D. 1966. Activation analysis as a tool for medical research. Nucleonics 24:54.

[3] Bowen, H., and I. Gibbons. 1963. Radioactivation analysis. Oxford Univ. Press, New York.

[4] Lenihan, J., and S. Thomson. 1954. Activation analysis. Academic Press, New York.

[5] Schütte, K. 1964. The biology of the trace element. J. B. Lippincott Co., Philadelphia.

[6] Lenihan, J. 1967. Nuclear activation analysis. *In* G. Hine (ed.), Instrumentation in nuclear medicine. Academic Press, New York. p. 309.

[7] Forshufvud, S., H. Smith, and A. Wassen. 1964. Napoleon's illness 1818–1821. Archiv für Toxicologie 20:210.

[8] Newton, D., J. Anderson, C. Battye, S. Osborn, and R. Tomlinson. 1969. Activation analysis *in vivo* using 5 MeV incident neutrons. Internat. J. Appl. Radiat. Isotopes 20:61.

[9] Cohn, S., C. Dombrowski, and R. Fairchild. 1970. *In vivo* neutron activation analysis of calcium in man. Internat. J. Appl. Radiat. Isotopes 21:127.

18

Statistics of Radioactivity Measurements

The moment of decay of a particular radioactive nucleus is not predictable and is not influenced by the history of the nucleus (that is, the period of time over which the nucleus has not decayed). Other factors, such as the environment or chemical and physical treatment of the nucleus, also do not influence the moment of decay. This unpredictability of the moment of decay is described as the randomness of radioactive decay. Because radioactive nuclei decay randomly, an understanding of probability and statistics is required for proper interpretation of data concerning radioactive samples.

Precision, Accuracy, and Bias

The deviation of a measurement from the "true value" for the quantity being measured is termed the *error* of the measurement. The error may be determinate, indeterminate, or a combination of both. Determinate error (sometimes called systematic error) is caused by one or more specific faults in the measurement process, and may be reduced by techniques such as using better instrumentation, planning and executing the measurement carefully, and correcting the measurement for identified sources of error. Indeterminate error (sometimes called random error) is inherent in the quantity being measured and in the measurement process, and cannot be reduced by eliminating extraneous factors which influence the measurement. In the measurement of radioactivity, the randomness of radioactive decay often is the greatest source of indeterminate error.

237

The range or "spread" of individual measurements in a series about an average value for the series is described as the precision of the measurements. Precision describes the reproducibility of the individual measurements and improves with reduction of the influence of indeterminate error upon the measurements. Precision does not describe the accuracy of the measurements; the measurements are accurate only if they agree with the "true value" for the quantity being measured. To achieve accuracy, the influence of both indeterminate and determinate error must be negligible. The influence of determinate error is described as the bias of the measurements. Precision, accuracy, and bias are distinguished in Figure 18-1.

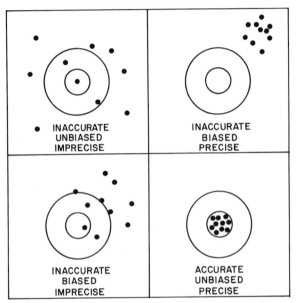

Figure 18-1   Precision, bias, and accuracy. [From W. Hendee. 1970. Medical radiation physics. Year Book Medical Publishers, Chicago. p. 369.]

Frequency Distribution of Radioactivity Measurements

From a large number of measurements of the decay rate of a long-lived radioactive sample, a graph may be constructed illustrating the frequency of obtaining a particular measurement as a function of the value of the measurement. The graph will resemble the curve in Figure 18-2. From the graph, the probability of obtaining a particular measurement or "count" may be estimated as the number of times the particular

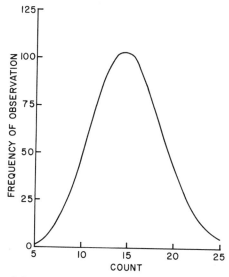

Figure 18-2 A Poisson distribution for a series of measurements with a mean of 15 for the series.

count was obtained, divided by the total number of times the sample was counted. This probability is expressed as

$$p_n = \frac{r^n e^{-r}}{n!}$$

where p_n is the probability of obtaining a count n, and r is the true average count for the sample. The term $n!$ (n-factorial) equals $(n)(n-1)(n-2)\dots$ $(2)(1)$. The probability p_n is termed the *Poisson probability density*, and radioactive decay is said to follow a Poisson probability law.

The true average count or "true mean" r for the sample cannot be measured; consequently, the average count or "estimated mean" $\hat{r}$ for a series of measurements usually is assumed to equal the true mean. With this assumption, the expression for the Poisson probability density p_n may be written

$$p_n = \frac{\hat{r}^n e^{-\hat{r}}}{n!}$$

Example 18-1
What is the probability of measuring a count of 10 for a radioactive sample when the average count is 12?

$$p_n = \frac{\hat{r}^n e^{-\hat{r}}}{n!}$$

$$p_{10} = \frac{(12)^{10}e^{-12}}{10!}$$

$$= \frac{(61.9 \times 10^9)\,(6.13 \times 10^{-6})}{36.3 \times 10^5}$$

$$= 0.105$$

The probability is 0.105 (or about 10%) that a count of 10 will be obtained when the average count is 12.

If a large number of counts is accumulated for a radioactive sample, then the curve of frequency of observation versus count (Figure 18-2) resembles closely the curve for a Gaussian or normal probability density function. The approach of a Poisson to a normal probability density curve is illustrated in Figure 18-3.

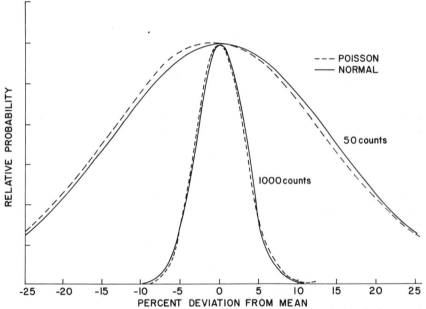

Figure 18-3 Relative probability of observing a particular count, for average counts of 50 and 1000. ——————, normal probability density functions; --------, Poisson probability density functions. Curves have been normalized to equal heights at the mean.

Standard Deviation

The precision or reproducibility of individual measurements of the activity of a radioactive sample may be described by the standard deviation of the measurements. The standard deviation σ is

$$\sigma = \sqrt{r}$$

where r is the true mean. If the true mean is estimated as the average value $\hat{r}$ for the measurements, then the estimated standard deviation $\hat{\sigma}$ is

$$\hat{\sigma} = \sqrt{\hat{r}}$$

If only one count n is obtained for the sample, then the standard deviation may be estimated as $\sqrt{n}$ by assuming that the single count represents the true mean.

In a normal distribution, 68.3% of all measured values fall within the limits of one standard deviation on either side of the mean. Within two standard deviations, 95.5% of all observations are enclosed; limits of three standard deviations encompass 99.7% of the observations.

For a series of measurements, the standard deviation of the measurements may be computed with the expression

$$\sigma = \sqrt{\frac{\sum\limits_{i=1}^{N} (n_i - r)^2}{N - 1}}$$

In this equation, N is the number of measurements from which the sample mean r is determined, and n_i represents the individual measurements.

Example 18-2
Determine the standard deviation for the series of measurements below.

Observation	n_i	$n_i - r$	$(n_i - r)^2$
1	48	−4	16
2	47	−5	25
3	61	+9	81
4	43	−9	81
5	57	+5	25
6	47	−5	25
7	57	+5	25
8	54	+2	4
9	49	−3	9
10	52	0	0
$\sum\limits_{i=1}^{N} n_i = 515$		$\sum\limits_{i=1}^{N} (n_i - r)^2 = 291$	

$$r = \frac{\sum\limits_{i=1}^{N} n_i}{N}$$

$$= \frac{515}{10}$$

$$\simeq 52$$

$$\sigma = \sqrt{\frac{\sum\limits_{i=1}^{N} (n_i - r)^2}{N - 1}}$$

$$= \sqrt{\frac{291}{10 - 1}}$$

$$\simeq 6$$

The mean for the sample may be expressed as

$$52 \pm 6$$

where 6 is understood to represent the standard deviation of the individual measurements.

The standard deviation σ_c of a count rate is

$$\sigma_c = \frac{\sigma}{t}$$

where σ is the standard deviation of the total accumulated count and t is the time over which the count was accumulated. Since

$$\hat{\sigma} = \sqrt{n}$$

the standard deviation of a count rate may be estimated as

$$\hat{\sigma}_c = \frac{\sqrt{n}}{t}$$

The count n is the average count rate c multiplied by the time t.

$$\hat{\sigma}_c = \frac{\sqrt{ct}}{t}$$

$$= \sqrt{\frac{c}{t}}$$

TABLE 18-1   *Rules for Arithmetic Operations Involving Numbers A and B*

Arithmetic operation	First number	Second number	Result $\pm$ standard deviation
Addition	$(A \pm \sigma_A)$	$+ (B \pm \sigma_B)$	$(A + B) \pm \sqrt{\sigma_A{}^2 + \sigma_B{}^2}$
Subtraction	$(A \pm \sigma_A)$	$- (B \pm \sigma_B)$	$(A - B) \pm \sqrt{\sigma_A{}^2 + \sigma_B{}^2}$
Multiplication	$(A \pm \sigma_A)$	$\times (B \pm \sigma_B)$	$(AB)[1 \pm \sqrt{(\sigma_A/A)^2 + (\sigma_B/B)^2}]$
Division	$(A \pm \sigma_A)$	$\div (B \pm \sigma_B)$	$(A/B)[1 \pm \sqrt{(\sigma_A/A)^2 + (\sigma_B/B)^2}]$

That is, the estimated standard deviation of a count rate is the square root of the count rate divided by the counting time.

If σ_A and σ_B denote the standard deviations of numbers A and B, then arithmetic operations involving A and B may be computed with the expressions in Table 18-1.

Example 18-3
A 2 min count of a sample yields 3 124 counts and a 2 min background count provides 1605 counts. Compute the net count rate and its standard deviation.
Count rate for sample (plus background) = 3124/2 min = 1562/min.
Estimated standard deviation $\hat{\sigma}_c$ of sample (plus background) count rate is

$$\hat{\sigma}_c = \sqrt{\frac{c}{t}}$$
$$= \sqrt{\frac{1562/\text{min}}{2 \text{ min}}}$$
$$= 28/\text{min}$$

Count rate for background = 1605/2 min = 802/min.
Estimated standard deviation $\hat{\sigma}_b$ of background count rate is

$$\hat{\sigma}_b = \sqrt{\frac{c}{t}}$$
$$= \sqrt{\frac{802/\text{min}}{2 \text{ min}}}$$
$$= 20/\text{min}$$

$$\text{net sample count rate} = (A - B) \pm \sqrt{\sigma_A{}^2 + \sigma_B{}^2}$$
$$= (1562 - 802) \pm \sqrt{(28)^2 + (20)^2}$$
$$= 760 \pm 34/\text{min}$$

The fractional standard deviation σ/r is the standard deviation σ for measurements in a series divided by the mean r for the series.
Since $\sigma = \sqrt{r}$, where r is the true mean,

$$\sigma/r = \frac{\sqrt{r}}{r}$$
$$= \frac{1}{\sqrt{r}}$$

For example, σ is 100 for a true mean of 10,000, and the fractional standard deviation is $100/10,000 = 0.01$. The percent standard deviation % σ is the fractional standard deviation multiplied by 100. In the example above, % $\sigma = 1\%$. The percent estimated standard deviation (% $\hat{\sigma}_c$) of a count rate c is

$$\% \hat{\sigma}_c = 100\left(\frac{\hat{\sigma}_c}{c}\right)$$

$$= \frac{100}{c}\sqrt{\frac{c}{t}}$$

$$= 100\sqrt{\frac{1}{ct}}$$

Useful Statistical Tests

A number of statistical tests may be applied to measurements of radioactive samples. A few of these tests are described briefly in this section.

Students' t Test. This test is a method for evaluating the statistical significance of the difference between two measurements. The t-value for two measurements n_1 and n_2 is

$$t\text{-value} = \frac{|n_1 - n_2|}{\sqrt{\sigma_1^2 + \sigma_2^2}}$$

where σ_1 and σ_2 are the standard deviations for the measurements n_1 and n_2, and the vertical bars enclosing $n_1 - n_2$ indicate that the positive (absolute) value of the quantity $n_1 - n_2$ should be used for the computation. After the t-value has been computed, the corresponding probability p is found in Table 18-2. The value of p is the probability that the difference in the numbers n_1 and n_2 is due to the random nature of radioactive decay and is not a real difference between dissimilar samples.

Example 18-4
Two 1 ml samples of blood are withdrawn from two animals involved in a radiotracer experiment. The sample from the first animal provided 924 ± 33 counts/min, and the sample from the second animal yielded 973 ± 37 counts/min. Is the difference significant?

$$t\text{-value} = \frac{|n_1 - n_2|}{\sqrt{\sigma_1^2 + \sigma_2^2}}$$

$$= \frac{|924 - 973|}{\sqrt{(33)^2 + (37)^2}}$$

$$= 0.99$$

From Table 18-2, the probability p is 0.322 (or 32.2%) that the difference is attributable to random fluctuation of the count rate for similar samples. The probability is $1.0 - 0.322 = 0.678$ that the difference between the samples is significant. In almost any situation, this difference between count rates would not be considered significant.

TABLE 18-2 *Cumulative   Normal   Frequency Distribution*

t-Value	p	t-Value	p
0.0	1.000	2.5	0.0124
0.1	0.920	2.6	0.0093
0.2	0.841	2.7	0.0069
0.3	0.764	2.8	0.0051
0.4	0.689	2.9	0.0037
0.5	0.617	3.0	0.00270
0.6	0.548	3.1	0.00194
0.7	0.483	3.2	0.00136
0.8	0.423	3.3	0.00096
0.9	0.368	3.4	0.00068
1.0	0.317	3.5	0.00046
1.1	0.272	3.6	0.00032
1.2	0.230	3.7	0.00022
1.3	0.194	3.8	0.00014
1.4	0.162	3.9	0.00010
1.5	0.134	4.0	0.0000634
1.6	0.110	4.1	0.0000414
1.7	0.090	4.2	0.0000266
1.8	0.072	4.3	0.0000170
1.9	0.060	4.4	0.0000108
2.0	0.046	4.5	0.0000068
2.1	0.036	4.6	0.0000042
2.2	0.028	4.7	0.0000026
2.3	0.022	4.8	0.0000016
2.4	0.016	4.9	0.0000100

Efficient Distribution of Counting Time.   If the time for counting sample and background is distributed as shown here, then the standard deviation of the net sample count rate is reduced to a minimum.

$$\frac{t_{s+b}}{t_b} = \sqrt{\frac{c_{s+b}}{c_b}}$$

t_{s+b} = counting time for sample (including background)

t_b = counting time for background

c_{s+b} = count rate for sample (including background)

c_b = count rate for background

Example 18-4

The count rate for a sample is estimated as 1624 cpm, including background. The background count rate is about 170 cpm. If a total counting time of 10 min is available, what counting time should be used for the sample and what counting time should be allotted to measurement of background?

$$\frac{t_{s+b}}{t_b} = \sqrt{\frac{c_{s+b}}{c_b}}$$
$$= \sqrt{\frac{1624}{170}}$$
$$= 3.1$$

Also, $t_{s+b} + t_b = 10$ min

Hence, $3.1\, t_b + t_b = 10$ min

$$t_b = 2.4 \text{ min}$$
$$t_{s+b} = (10.0 - 2.4) \text{ min}$$
$$= 7.6 \text{ min}$$

The sample might be counted for 7.5 min and the background for 2.5 min.

Rejection of Individual Measurements

Occasionally, a choice must be made between obtaining a single long measurement of sample activity and obtaining a series of shorter measurements from which the activity of the sample could be computed as the mean of the individual measurements. For samples which furnish a relatively high count rate, the two approaches yield essentially the same precision. For samples of low count rate, the second method is preferred because it tends to provide slightly greater precision.

A single measurement always should be rejected if an external influence on the measurement has been identified. Even though no external influence has been identified, a single measurement in a series may differ widely from the mean for the series. To decide whether the measurement reflects statistical fluctuation in decay of the sample or the presence of an unidentified source of determinate error, Chauvenet's criterion may be used. This criterion states that a single measurement may be rejected if the probability of its occurrence is equal to, or less than, $1/2N$, where N is the number of measurements in the series. The computation of the probability of occurrence of a measurement is time consuming; consequently, an alternate approach to the application of Chauvenet's criterion usually is followed. This approach requires the computation of $|n - r|/\sigma$, where n is the measurement, r is the mean for the series, and σ is the standard deviation of individual measurements in the series. The measurement may be rejected if $|n - r|/\sigma$ exceeds the value in Table 18-3 for the number of measurements in the series.

TABLE 18-3   *Chauvenet's Criterion*

Number of measurements N	$\|n-r\|/\sigma$	Number of measurements N	$\|n-r\|/\sigma$
2	1.15	15	2.13
3	1.38	20	2.24
4	1.54	25	2.35
5	1.65	30	2.40
6	1.73	35	2.45
7	1.80	40	2.50
8	1.86	50	2.58
9	1.91	75	2.71
10	1.96	100	2.81
12	2.04	200	3.02

Example 18-5

Five 1 min counts of a sample yielded the following results:

$$990, 1084, 1029, 996, 1005$$

Should the count of 1084 be discarded?

$$\text{mean} = \frac{\sum\limits_{i=1}^{N} n_i}{N}$$
$$= \frac{5104}{5}$$
$$= 1021$$

$$\text{standard deviation } \sigma = \sqrt{\frac{\sum\limits_{i=1}^{N} (n_i - r)^2}{N-1}}$$
$$= \sqrt{\frac{5875}{4}}$$
$$= 38$$

For the 1084 count

$$\frac{|n-r|}{\sigma} = \frac{|1084 - 1021|}{38}$$
$$= \frac{63}{38}$$
$$= 1.66$$

Since a value of 1.66 for $|n - r|/\sigma$ is just greater than the value of 1.65 listed for five measurements in Table 18-3, the count of 1084 may be rejected.

Precision of Ratemeter Measurements

The estimated standard deviation $\hat{\sigma}_c$ of a ratemeter reading c in counts per second is

$$\hat{\sigma}_c = \sqrt{\frac{c}{2RC}}$$

where RC is the time constant in seconds for the ratemeter. The percent estimated standard deviation $\% \, \hat{\sigma}_c$ is

$$\% \, \hat{\sigma}_c = 100 \sqrt{\frac{1}{2cRC}}$$

The $\% \, \hat{\sigma}_c$ is reduced if the count rate is increased or if it is averaged in the ratemeter over a longer interval of time.

Example 18-6

A ratemeter with a time constant of 0.5 sec displays an equilibrium count rate of 1000 cpm. What are the estimated standard deviation $\hat{\sigma}_c$ and the percent estimated standard deviation $\% \, \hat{\sigma}_c$?

$$\hat{\sigma}_c = \sqrt{\frac{c}{2(RC)}}$$
$$= \sqrt{\frac{1000/60}{2(0.5 \text{ sec})}}$$
$$\approx 4/\text{sec}$$

$$\% \, \hat{\sigma}_c = 100 \sqrt{\frac{1}{2c(RC)}}$$
$$= 100 \sqrt{\frac{1}{2(1000/60)(0.5)}}$$
$$= 24$$

PROBLEMS

1. 3240 counts were accumulated for a sample counted for 10 min in the presence of background. Background provided 812 counts during a counting interval of 8 min. What are:

(a) The estimated standard deviation and the percent estimated standard deviation of the sample count rate uncorrected for background?

(b) The estimated standard deviation and the percent estimated standard deviation of the background count rate?

(c) The estimated standard deviation and the percent estimated standard deviation of the sample count rate corrected for background?

2. A 2 min count of 3130 was measured for a radioactive sample. A second sample furnished 4529 counts in 3 min. Is the difference in count rate between the two samples significant?

3. For time constants of 0.5 sec and 5 sec, compare the percent estimated standard deviation for a ratemeter reading of 1500 cpm.

4. Five 1 min observations of sample activity yielded the following counts per minute: 124, 175, 109, 105, 111. Can any of the observations be rejected by Chauvenet's criterion?

19

Quantitative Measurement of Radioactivity

The count rate for a radioactive sample reflects the rate of decay of atoms in the sample. To determine the activity of a sample from the measured count rate, the influence of determinate errors on the count rate must be estimated. The computation of activity from count rate may be stated as

$$A = \frac{c_{s+b}}{(1 - c_{s+b}T)EfGBf_c f_w f_s} - c_b$$

where corrections to the background count rate are assumed to be negligible. In this expression,

A = activity in disintegrations per minute
c_{s+b} = (sample plus background) count rate in counts per minute
c_b = background count rate in counts per minute
T = resolving time in minutes
E = detector efficiency
f = fractional emission of source
G = geometry correction
B = backscatter correction
f_c = sidescatter correction
f_w = correction for attenuation in detector window, air, sample covering, etc.
f_s = correction for sample self-absorption

These corrections are discussed in the following sections.

Background Count Rate

A counting system used to detect radiation from a radioactive sample indicates the presence of radiation when the sample is removed. The residual radiation is termed *background radiation* and can be attributed to a variety of sources, including (1) cosmic radiation; (2) radioactive materials such as radium, ^{14}C, and ^{40}K in the earth, in the human body, and in construction materials; (3) radioactive materials stored or used near the counting system; (4) radioactive materials used in devices such as watches and instrument dials; (5) radioactive contamination of the counting equipment, laboratory benches and apparatus, etc.; and (6) radioactive fallout (the contribution from this source usually is 1% or less). To reduce the background count rate, radiation detectors usually are surrounded by a lead shield. Other methods to reduce the background count rate include careful technique, the use of a pulse-height analyzer, and the inclusion of a coincidence or anticoincidence circuit in the counting system.

Resolving Time

The resolving time of a radiation detector is the time required between successive interactions in a radiation detector for the interactions to be recorded as separate events. Compared to other detectors (Table 19-1), GM tubes have long resolving times (200–300 μsec). The resolving time of a GM counter may be estimated with fair accuracy by the paired source method[1]; more accurate measurements may be obtained with an oscilloscope. The count rate c_0, corrected for the loss of counts (the coincidence loss) caused by the resolving time of the instrument, is

$$c_0 = \frac{c}{1 - cT}$$

TABLE 19-1 *Typical Output Signals, Amplification Factors, and Resolving Times for Various Detectors*

Detector	Output signal	Amplification factor	Resolving time
Pulse ionization chamber	0–1 mV	1	1 μsec
Proportional counter	0–100 mV	10^2–10^4	5–50 μsec
G–M tube	0–10 V	10^6–10^8	10–1000 μsec
NaI(Tl) scintillation detector	0–2 V	10^6	< 1 μsec
Liquid scintillation cocktail	0–100 mV	10^6	~ 1 nsec
Semiconductor detector	0–25 mV	1	negligible

where c is the measured count rate and T is the resolving time. For a typical GM tube, the coincidence loss is about 0.5%/1000 cpm.

Example 19-1
What is the count rate corrected for coincidence loss for a measured count rate of 3500 cpm furnished by a GM detector with a resolving time of 300 μsec?

$$c_0 = \frac{c}{1 - cT}$$

$$= \frac{3500/\text{min}}{1 - (3500/\text{min})(300 \times 10^{-6}\,\text{sec})(1/60\,\text{sec}/\text{min})}$$

$$= 3562/\text{min}$$

Detector Efficiency

The ratio of the number of particles or photons interacting in a detector to the number entering the detector is termed the *efficiency* of the detector. The efficiency of most GM and proportional detectors is close to 1 for α and β particles and perhaps 0.01 for high-energy x- and γ-rays. The efficiency of a NaI(Tl) or a semiconductor detector for γ-rays varies with the size of the detector and the energy of the γ-rays.

Fractional Emission

The fractional emission of a radioactive source is the fractional number of decays which results in release of the radiation to be detected. For example, about 93.5% of all decays of ^{137}Cs involves an isomeric transition of 0.66 MeV. This transition occurs by γ emission 90% of the time and by internal conversion 10% of the time. The fractional emission of γ-rays from a ^{137}Cs source is $(0.935)(0.9) = 0.84$.

Detector Geometry

The geometry correction is the ratio of the number of particles or photons emitted from a source in the direction of a detector to the total number of emitted particles or photons. Usually, the detector is described as subtending a certain solid angle Ω measured in steradians. Since a sphere subtends a solid angle of 4π steradians, the geometry correction G is

$$G = \frac{\Omega}{4\pi}$$

The expressions "2π geometry" and "4π geometry" describe situations where half or all of the radiation emitted by a source is intercepted by the detector. For a point radioactive source and a cylindrical detector with an entrance window at one end (Figure 19-1), the geometry correction G varies with the radius r of the detector and the distance h between the detector and source:

$$G = 0.5\left(1 - \frac{h}{\sqrt{h^2 + r^2}}\right)$$

For a disk source of radius a, the geometry correction G is:

$$G = 0.5\left(1 - \frac{h}{\sqrt{h^2 + r^2}}\right) - \frac{3}{16}\left(\frac{ar}{h^2}\right)^2\left(\frac{h}{\sqrt{h^2 + r^2}}\right)^5$$

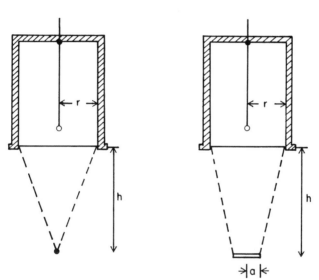

Figure 19-1 Geometric relationship between a cylindrical detector and a point (left) and a disk (right) (radius = a) source of radioactive material.

Scattering

Radiation emitted from a source in any direction may be scattered toward a detector during interactions of the radiation with surrounding materials (for example, the sample support or shielding around source and detector). The scattered radiation is termed *backscatter* or *sidescatter*. If the backscattered radiation originates entirely within the sample support (Figure 19-2), then the percent backscatter (% B) and

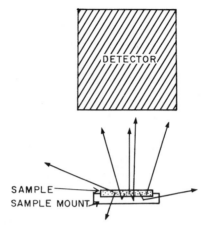

Figure 19-2 Backscattered radiation from the support for a radioactive sample.

backscatter factor B are:

$$\% \, B = \frac{(\text{counts with support present}) - (\text{counts with support removed})}{\text{counts with support removed}} \times 100$$

$$B = \frac{\text{counts with support present}}{\text{counts with support removed}}$$

The backscatter factor is related to the percent backscatter by

$$\% \, B = 100(B-1)$$

For β particles, the backscatter factor increases with the atomic number of the sample support and, initially, with the thickness of the support. Saturation thickness is achieved when a further increase in support thickness does not contribute additional backscatter (Figure 19-3). For β particles, saturation thickness equals about 3/10 of the range of β particles in the support material. Backscattering may be reduced by using a support material of low atomic number (Figure 19-3). For β counting, for example, aluminum planchets often are preferred over copper or steel because the backscatter factor is lower for aluminum. For β-emitting substances mounted on supports of saturation thickness, the backscatter factor is almost independent of the energy of the β particles.

Sidescatter may be reduced by positioning all scattering material (for example, shielding around source and detector) far from the path between

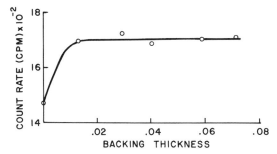

Figure 19-3 Backscattering of ^{204}Tl β particles from aluminum sample supports of different thicknesses.

source and detector. A sleeve of plastic, aluminum, or other low-Z material inside the detector-source shield also reduces the sidescatter.

Air and Window Absorption

Radiation is attenuated by any material between the radioactive source and the sensitive volume of the detector. Usually, at least three media are encountered by radiation moving from a source toward a detector. These media are: (1) the covering for the radioactive source; (2) air between the source and detector; and (3) the entrance window of the detector.

Self-Absorption

In any radioactive sample, some of the radiation is attenuated within the sample itself. This loss of radiation is termed *self-absorption*. Shown in Figure 19-4 is the change in count rate accompanying an increase in activity of a β-emitting sample with no change in sample volume. The data describe a straight line, because the fractional loss of counts by self-absorption remains constant. However, if the volume of the sample increases with no change in total activity, then the count rate for the sample decreases because an increasing fraction of the radiation is absorbed (Figure 19-5).

In Figure 19-6, the count rate is illustrated for a sample with increasing volume and constant specific (rather than total) activity. After the sample reaches an infinite thickness, the count rate remains constant because the increased attenuation of radiation by the sample compensates for the increased activity of the sample as the thickness of the sample is made greater.

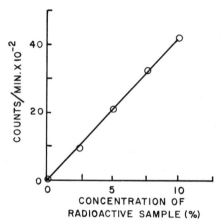

Figure 19-4   Count rate for a sample with increasing activity and constant volume.

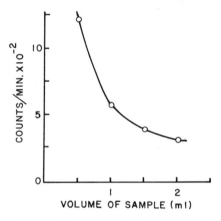

Figure 19-5   Count rate for a sample with increasing volume and constant specific activity.

A self-absorption correction curve for samples of a particular radio-active substance may be obtained by preparing a series of samples with identical specific activities but different volumes and, consequently, different total activities. The count rate for each sample is measured and divided by the mass or volume of the sample. The resulting data are plotted as a function of sample mass or volume. By extrapolating the curve to a mass or volume of 0, a value is obtained for the specific activity of the samples corrected for self-absorption (Figure 19-7). For any particular sample in the series, the count rate corrected for self-absorption

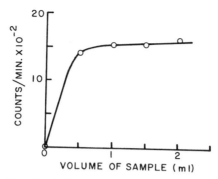

Figure 19-6 Count rate for a sample with constant specific activity but increasing volume and total activity.

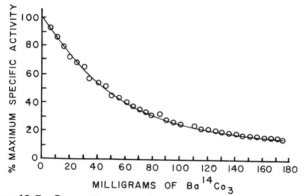

Figure 19-7 Correction curve for self-absorption of samples of ^{14}C-labeled $BaCO_3$. [From C. Wang and D. Willis. 1965. Radiotracer methodology in biological science. Prentice-Hall, Englewood Cliffs, N.J. p. 205.]

may be determined by multiplying the mass or volume of the sample by the corrected specific activity.

Absolute and Secondary Standards

An absolute standard (sometimes called a primary standard) is a radioactive sample with a disintegration rate which has been determined without reference to another standard. To eliminate the necessity for restandardization at frequent intervals, the half-life of an absolute standard should be relatively long. The decay scheme for an absolute standard must be known accurately.

Responsibility for maintenance of international absolute standards of radioactivity resides with the International Commission on Radiological Units and Measurements (ICRU). The international reference standard of radioactivity is a 22.23 mg source of radium chloride prepared in 1934.

Secondary standards are samples obtainable from national standardizing laboratories (the National Physical Laboratory in Great Britain and the National Bureau of Standards, NBS, in the United States). Each of these samples has a disintegration rate which has been determined by comparison to an absolute standard. Secondary standards are relatively inexpensive and are useful for a variety of purposes in a counting laboratory. However, they must be handled and used carefully to maintain the accuracy of the standardization. For example, the following comments should be considered when solid and liquid radioactive standards are used[2].

(1) Standard solutions are prepared so that the radioactive component remains in solution and does not attach to the walls of the container. If the standard must be diluted, then a diluent should be used which does not initiate deposition of the radioactive component on the walls or bottom of the vessel.

(2) Bacterial growths can absorb the radioactive component of a standard solution; consequently, bacterial growths are inhibited by heat-sterilization or by addition of a bacteriostat to a standard solution. Similar precautions should be followed if a standard solution is diluted.

(3) Because evaporation of a standard solution can change its specific activity, a standard always should be stored in a closed, air-tight container. Plastic containers should not be used, because the solution may evaporate through the walls of the vessel.

(4) The container for a standard sample should be chosen with the chemical form of the standard in mind. For example, a standard dissolved in hydrochloric acid will dissolve an aluminum planchet.

(5) For most standard solutions and solid sources, the activity per milliliter or per gram is specified. To determine the types and amounts of radiation from a standard, the decay scheme for the nuclide must be consulted, and absorption and scattering of the radiation must be considered.

(6) The disintegration rate of a standard always must be corrected for radioactive decay occurring since certification of the standard. Also, the effects of long-lived and short-lived impurities which may be present in the standard must be considered, as well as the buildup and decay of radioactive daughters in the standard.

(7) Variations in size, shape, and composition between a standard and samples of unknown activity may contribute different conditions for absorption and scattering. These differences must be considered when estimates are made of the activity of the unknown samples.

Coincidence Counting

If a sample contains a nuclide which decays by emission of two or more particles or photons, then the activity of the sample may be determined by coincidence counting. For example, a nuclide which decays by emission of a β particle and a γ-ray photon may be counted by operating a detector for the β particles in coincidence with another detector sensitive only to γ-rays. The β count rate C_β is

$$C_\beta = AE_\beta$$

where A is the activity of the source and E_β is the overall efficiency for detection of β particles (that is, E_β is a summation of all the terms discussed earlier in this chapter). The gamma count rate C_γ is

$$C_\gamma = AE_\gamma$$

where E_γ is the overall efficiency for γ counting. The count rate $C_{\beta\gamma}$ for the detectors operated in coincidence is

$$C_{\beta\gamma} = AE_\beta E_\gamma$$

From this expression, the β efficiency is

$$E_\beta = \frac{C_{\beta\gamma}}{AE_\gamma}$$

since $AE_\gamma = C_\gamma$

$$E_\beta = \frac{C_{\beta\gamma}}{C_\gamma}$$

The activity A of the sample is

$$A = \frac{C_\beta}{E_\beta}$$

and, with $C_{\beta\gamma}/C_\gamma$ substituted for E_β

$$A = \frac{C_\beta C_\gamma}{C_{\beta\gamma}}$$

This expression for sample activity is independent of the counting efficiencies E_β and E_γ.

In actual practice, a number of corrections must be applied to coincidence counting data before the activity is computed. For example, internal conversion processes, the sensitivity of the β detector to γ-rays, and the presence of impurities and daughter nuclides in the sample must be considered. The proper interpretation of coincidence counting data can be quite difficult, particularly if the nuclide exhibits multiple pathways for decay. By measurement of x-rays or Auger electrons in coincidence with γ-rays, sample activities can be determined for nuclides which decay by electron capture and γ emission. Similar techniques can be applied to nuclides which emit two or more γ-ray photons in cascade. Anticoincidence counting techniques can be used for nuclides which emit particulate radiations followed sometimes, but not always, by γ-ray photons.

PROBLEMS

1. A GM detector which is 90% efficient for counting β particles subtends a solid angle of 0.3 steradians for a sample of ^{204}Tl ($T_{1/2} = 3.8$ yr). The scaler registers 180 cpm corrected for background. Assuming that corrections for backscatter, sidescatter, air, window and self-absorption all equal 1.0, and that coincidence losses are negligible, compute the number of ^{204}Tl atoms in the sample.

2. A GM detector has a resolving time of 250 μsec. What is the count rate corrected for coincidence loss if the measured count rate is 8,000 cpm? What is the maximum observable count rate for this detector for a coincidence loss which does not exceed 1%?

3. The following data were recorded for identical samples:
   "weightless" Mylar mount   1888
   silver backing   3108
   Compute the percent backscatter and the backscatter factor.

4. Explain why there is a saturation or infinite thickness for curves of backscattering (count rate versus thickness of source support) and self-absorption (count rate versus weight of sample for samples of constant specific activity).

5. The following data were obtained by counting a series of weighted fractions of a radioactive sample:

mass, mg	1	2	4	6	8	10	15	20	25	30	35
cpm, net	67	124	210	285	355	385	410	430	448	458	465

Plot a self-absorption curve and determine the specific count rate (cpm/mg) for the sample corrected for self-absorption.

6. A radioactive sample of short half-life will be counted once each day over a period of several days. Explain how a correction can be determined for variations in the observed count caused by fluctuations in the sensitivity of the counting system.

7. An ^{131}I standard solution obtained from the NBS was accompanied by the following information: assay of ^{131}I at 8:00 AM, March 15 = 4.13×10^4 dps/ml $\pm 2\%$. On March 30 at 8:00 AM, a sample of ^{131}I was counted. This sample had been obtained from a shipment of ^{131}I in the following manner: 20λ ($1\lambda = 1$ μliter) of the shipment were diluted in a volumetric flask to 25 ml; 100 λ of this solution were diluted to 10 ml; 50λ of this solution were withdrawn, mounted on Mylar, dried, and counted. The dried sample provided a count rate of 4150 cpm corrected for background. At the same time (8:00 AM, March 30) a 50 λ sample of the NBS standard ^{131}I solution was mounted, dried and counted in the same manner. The count rate for the standard sample was 2332 cpm corrected for background.

(a) What was the specific activity (mCi/ml) of the ^{131}I shipment at the time the sample from the shipment was counted?

(b) What was the total correction factor between count rate and activity for the counting system used to count the ^{131}I samples?

REFERENCES

[1] Chase, G., and J. Rabinowitz. 1967. Principles of radioisotope methodology, 3rd ed. Burgess Publishing Co., Minneapolis. p. 114.

[2] Wilson, B. J. (ed.). The radiochemical manual, 2nd ed. The Radiochemical Centre, Amersham, England. p. 118.

20

Dynamic Function Studies

Radioactive isotopes are used in various ways for measuring physiologic compartments and studying dynamic processes in the body. These uses may be separated into four categories: (1) isotope dilution studies; (2) isotope transfer studies; (3) isotope uptake and disappearance studies; and (4) isotope distribution studies. Three of these categories are discussed in this chapter, and the fourth is discussed in Chapter 21.

Isotope Dilution Studies

For a radioactive solution of volume V_i in milliliters and specific count rate C_i in counts per minute per milliliter, the total count rate is

total count rate (cpm) = specific count rate C_i (cpm/ml) × volume V_i(ml)

If this solution is mixed with a larger volume V_u of nonradioactive solution, then the specific count rate will be reduced to C_f (counts per minute per milliliter). If none of the radioactive solution is lost during mixing, then the total count rate may also be written

total count rate (cpm) = specific count rate C_f (cpm/ml) × volume V_f (ml)

where V_f is the volume of the mixture. Consequently,

$$C_f V_f = C_i V_i$$

and the final volume may be determined if the initial volume and specific count rates of the initial and final solutions are known.

$$V_f = \frac{C_i V_i}{C_f}$$

Since $C_i V_i$ = the total count rate of the solution before mixing, the final volume V_f may also be described as

$$V_f = \frac{\text{total count rate before mixing}}{C_f}$$

The volume V_u of the nonradioactive solution is

$$V_u = V_f - V_i$$

To measure a volume accurately by the dilution technique, all count rates must be corrected for background and for counting losses such as those attributed to coincidence loss and escape of radioactive material during mixing of the radioactive and nonradioactive solutions.

Dilution measurements are used to determine the size of physiologic spaces in the body and to measure the volume of blood, red cells, and plasma comprising the vascular pool. For example, tritiated water may be used to measure total body water, and radioactive nuclides of potassium, sodium, and chlorine (^{82}Br used often in place of chlorine) are used to measure total body spaces for these elements. The measurement of plasma, red blood cell, and blood volume by the dilution technique is illustrated in Example 20-1.

Example 20-1

A patient was injected intravenously with 5 μCi of serum albumin labeled with ^{131}I. A second 5 μCi sample of the albumin solution was diluted to a volume V_s of 10,000 ml. A 3 ml aliquot of this solution provided a count rate of 1100 cpm. A 3 ml sample of the patient's plasma provided 3950 cpm. The background count rate was 140 cpm. What was the patient's plasma volume? If the venous hematocrit was 45%, what were the red blood cell (RBC) and blood volumes?

$$\text{plasma net specific count rate } C_p = \frac{(3950 - 140) \text{ cpm}}{3 \text{ ml}}$$

$$= 1270 \text{ cpm/ml}$$

$$\text{standard net specific count rate } C_s = \frac{(1100 - 140) \text{ cpm}}{3 \text{ ml}}$$

$$= 320 \text{ cpm/ml}$$

The total count rate for 5 μCi of albumin solution is

$$\text{total count rate} = C_s V_s$$
$$= (320 \text{ cpm/ml}) (10,000 \text{ ml})$$
$$= 3.20 \times 10^6 \text{ cpm}$$

This total count rate also represents the $5 \mu\text{Ci}$ albumin solution before injection into the patient. The final volume V_f of labeled albumin and patient's plasma is

$$V_f = \frac{\text{total count rate before mixing}}{C_p}$$

$$= \frac{3.20 \times 10^6 \text{ cpm}}{1270 \text{ cpm/ml}}$$

$$= 2520 \text{ ml}$$

The volume of labeled albumin delivered to the patient is negligible, and the patient's plasma volume also is 2520 ml.

Since the venous hematocrit is 0.45, the blood volume is

$$\text{blood volume} = \frac{2520 \text{ ml}}{1.0 - 0.45}$$

$$= 4580 \text{ ml}$$

and the volume of red blood cells is

$$\text{RBC volume} = (4580 \text{ ml})(0.45)$$

$$= 2060 \text{ ml}$$

An alternate method for determining the RBC volume involves incubation of a sample of blood for 20–40 min in a noncoagulating solution of ^{51}Cr as sodium chromate. After the red blood cells have incorporated the ^{51}Cr, part of the blood sample is reinjected into the patient. After 15–20 min, a sample of blood is withdrawn for counting. The RBC volume V_f is

$$V_f = \frac{C_i V_i}{C_f} \quad \text{where}$$

C_i = specific count rate of incubated RBC
V_i = volume of incubated RBC injected
C_f = specific count rate of patient's RBC after injection of labeled sample

In practice, a number of corrections are required in all methods for measuring plasma, red blood cell, and blood volume[1].

Rate of Movement Studies

The time required for circulation of the blood from one region of the body to another may be estimated by injecting a bolus of activity of an appropriate nuclide (such as ^{24}Na) into the vascular supply for one region and timing the arrival of the activity in the second region. Similarly, the transit time of blood among various compartments of the heart and lungs may be estimated by injecting a compound such as ^{131}I-labeled serum albumin and using external detectors positioned over the precordium and lungs to time the arrival of the labeled compound at selected locations. These data may aid in the diagnosis of pulmonary hypertension and abnormal cardiac output. Similar rate of movement studies have been developed for the liver and kidneys.

Time-Concentration Studies

Hippuran (sodium orthoiodohippurate) tagged with ^{131}I is used widely in studies of kidney function. The accumulation and release of the compound in the kidneys is monitored by a scintillation detector over each kidney. The resultant curves (termed a renogram) of count rate versus time depict the extraction of the labeled compound from the blood into each kidney and its subsequent release from each kidney into the bladder. In Figure 20-1, the retention of activity in the right kidney suggests the presence of an obstruction in the kidney or ureter which interferes with the release of urine into the bladder.

A variety of time-concentration studies have been devised for evaluation of the thyroid gland. These studies rely upon measurements of the rate of accumulation of iodine by the thyroid gland, the total amount of

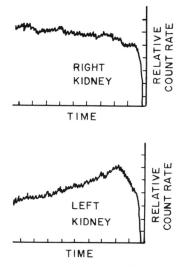

Figure 20-1   Renogram obtained with ^{131}I-Hippuran depicting a normal left kidney and a right kidney with increased retention of activity.

iodine accumulated in the thyroid over a period of time, or the rate of release of thyroid hormone from the gland. Test results are influenced not only by the functional state of the thyroid, but also by medications and drugs administered earlier to the patient, physiologic abnormalities such as malabsorption syndromes and renal disease, and also by external application of iodine.

In the most common measurement of thyroid function, 5–25 μCi of ^{131}I as NaI (or an equivalent dose of ^{123}I or ^{125}I) are administered orally, and the count rate over the thyroid gland is measured at selected times (usually 6 and 24 hr) with a NaI scintillation detector positioned at a prescribed distance from the gland. The percent uptake of iodine by the thyroid is determined by comparing the count rate for the gland to that obtained with a Lucite neck phantom counted in the same geometry and containing an amount of iodine equal to that administered to the patient (Example 20-2).

$$\text{uptake } (\%) = (100)\frac{c_t}{c_s}$$

where c_t = count rate for the thyroid and c_s = count rate for the standard in the neck phantom.

Radiation scattered toward the detector from extrathyroidal tissue may contribute to errors in the measurement of thyroid uptake. These errors may be reduced by using a pulse-height analyzer to reject pulses produced by scattered radiation, or by using a lead filter to estimate the contribution of extrathyroidal radiation to the count rate for the thyroid and neck phantom[2]. With this approach, the percent uptake of iodine in the thyroid is

$$\text{uptake } (\%) = (100)\frac{c_t - c_{tl}}{c_s - c_{sl}}$$

c_t = count rate for thyroid
c_{tl} = count rate for thyroid shielded by lead filter
c_s = count rate for standard in neck phantom
c_{sl} = count rate for standard in neck phantom shielded by lead filter

Example 20-2

The following count rates were recorded 24 hr after administration of 10 μCi of Na[131]I to a patient:

patient's thyroid	771 cpm
patient's thyroid shielded by lead filter	201 cpm
standard in neck phantom	4541 cpm
standard in neck phantom shielded by lead filter	87 cpm

The percent uptake was:

$$\text{uptake } (\%) = 100 \, \frac{c_t - c_{tl}}{c_s - c_{sl}}$$

$$= (100) \, \frac{771 - 201}{4541 - 87}$$

$$= 13$$

The range of percent uptakes which reflects normal or abnormal function of the thyroid may vary somewhat from one laboratory to another. Consequently, an uptake of 13% might indicate hypothyroidism in one laboratory and a normal thyroid in another.

Thyroid function may also be evaluated by measuring the activity excreted in the urine over a 24 or 48 hr period following administration of radioactive iodine. For a patient with a normal thyroid, from 35 to 65% of the administered radioactivity is excreted during the first 24 hr; values above this range imply hypothyroidism, and excreted amounts below 35% suggest hyperthyroidism. Difficulty in obtaining a complete urine specimen over a 24 hr period is the major disadvantage of this method.

The PBI (protein-bound iodine) conversion ratio also is indicative of thyroid function. The PBI conversion ratio is the ratio of iodine incorporated into the thyroid hormone thyroxine and bound to plasma protein divided by the total activity of iodine in the serum or plasma at a selected time (usually 24 hr) after administration of 25–100 μCi of ^{131}I. The iodine bound to plasma protein and that present in inorganic form in the plasma are separated by precipitating the protein with an agent such as trichloroacetic acid, or by extracting the inorganic iodine on an ion exchange column. The PBI conversion ratio, expressed as a percentage, is computed as

$$\text{PBI conversion ratio } (\%) = (100) \, \frac{\text{protein-bound radioiodine}}{\text{total plasma radioiodine}}$$

Example 20-3

The following counts were obtained for a PBI conversion ratio test. What was the conversion ratio?

count rate for 5 ml of plasma	816 cpm
count rate for protein-bound iodine in 5 ml of plasma	276 cpm
count rate for background	52 cpm

$$\text{PBI conversion ratio, } \% = (100) \frac{\text{protein-bound radioiodine}}{\text{total plasma radioiodine}}$$

$$= (100) \frac{276 - 52 \text{ cpm}}{816 - 52 \text{ cpm}}$$

$$= 29$$

Conversion ratios in the range of 12–45% generally are considered normal.

The hormones thyroxine and thyronine released by the thyroid into the blood are bound primarily to plasma proteins; however, red blood cells also attract the hormones. In a patient receiving radioactive iodine, little activity is captured by the red blood cells until the binding sites of the plasma proteins have become saturated with hormone. In 1957, Hamolsky, Stein and Freedberg[3] suggested that the amount of ^{131}I-labeled thyronine attached to red blood cells in a patient's blood reflects the number of binding sites available for labeled hormone on plasma proteins and, consequently, the concentration of nonradioactive thyroid hormones in the blood. The amount of ^{131}I-labeled thyronine attached to red blood cells can be estimated by measuring the activity of washed blood cells after incubation of a small sample of the patient's blood for a couple of hours in labeled thyronine. More recently, methods have been developed to determine the amount of radioactive thyroxine remaining unattached to either plasma proteins or red blood cells after incubation. This approach to evaluation of thyroid function is very popular, because the procedure is straightforward, the results are reliable and are not influenced greatly by exogenous iodine or iodides, and no radioactive material is administered to the patient. The procedure is outlined in Figure 20-2.

Absence of the intrinsic factor of Castle in the blood is associated with pernicious anemia and is the cause for poor absorption of vitamin B_{12} from the GI tract. In patients with pernicious anemia, the excretion of vitamin B_{12} into the urine is low. If vitamin B_{12} labeled with ^{57}Co or ^{58}Co (cyanocobalamin) is administered orally to a patient with pernicious anemia, the urinary excretion of radioactive cobalt over the following 24 hr will be 5% or less of the administered dose. The urinary excretion may be increased two to tenfold by oral administration of intrinsic factor. This increase is not observed for patients with sprue, idiopathic steatorrhea, or other malabsorption problems which also are accompanied by reduced absorption of vitamin B_{12}. This procedure for diagnosis of pernicious anemia usually is described as the Schilling test[4].

In certain types of hemolytic anemia, red blood cells are phagocytized at a greatly increased rate in the spleen, liver and bone marrow. The

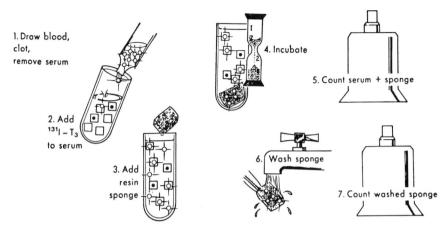

1. Draw blood, clot, remove serum
2. Add $^{131}I - T_3$ to serum
3. Add resin sponge
4. Incubate
5. Count serum + sponge
6. Wash sponge
7. Count washed sponge

Figure 20-2 Outline of labeled thyronine (T_3) test for evaluation of thyroid activity. [From E. Powsner and D. Raeside. 1971. Diagnostic nuclear medicine. Grune and Stratton, New York. p. 411, by permission.]

rate of phagocytosis occurring in a patient may be estimated by measuring the lifetime of red blood cells labeled with ^{51}Cr and injected into the patient. The cells are labeled by incubating a small sample of the patient's blood for 20–40 min in ^{51}Cr-labeled sodium chromate. The labeled red blood cells are reinjected and samples of the patient's blood are withdrawn and counted periodically over succeeding weeks. The time for the specific count rate of the blood to fall to half its initial value is the half-life of the red blood cells. For the normal patient, half-lives between 28 and 32 days usually are measured. Measurements must be corrected for decay of ^{51}Cr and for an elution loss of about 1% per day caused by detachment of ^{51}Cr from the cells.

The digestion and absorption of fat in the GI tract may be studied by blocking the patient's thyroid with Lugol's solution and administering ^{125}I- or ^{131}I-labeled triolein or oleic acid orally in combination with a high-fat meal. Blood samples are withdrawn and counted periodically for the next 24 hr to determine the portion of the administered activity present in the blood. The activity excreted in the feces also may be measured. In persons with malabsorption problems, plasma levels of activity are reduced and fecal levels are increased[4].

The functional capability of the liver may be evaluated by using an external detector to estimate the activity in the liver as a function of time after intravenous administration of ^{131}I-labeled rose bengal dye. Alternately, the rate of disappearance of the ^{131}I-labeled rose bengal dye from the blood may be measured. In a normal patient, the activity in the liver is greatest 15–20 min after injection of the dye, and decreases gradually thereafter as the dye is excreted. Reduced activity in the liver is noted for patients with diseased livers; in patients with a biliary obstruction, the uptake is normal but the rate of excretion is reduced.

Iron deficiency anemia may be detected by oral administration of ^{59}Fe and subsequent measurement of the amount of ^{59}Fe excreted in the feces. Radioactive iron also may be administered intravenously to determine the rate at which ^{59}Fe is cleared from the plasma and incorporated into red blood cells. In aplastic anemia, the clearance rate and incorporation of ^{59}Fe are reduced, whereas in iron deficiency anemia the rate of clearance and incorporation are rapid. In hemolytic anemia, the clearance is rapid and the uptake into red blood cells may be increased, and in hemochromatosis the plasma clearance time is lengthened and the uptake into red blood cells is reduced[1].

PROBLEMS

1. A patient received an intravenous injection of 3 μCi of serum albumin labeled with ^{131}I. From a second 3 μCi of labeled serum albumin diluted to 3000 ml, a 5 ml aliquot furnished a count rate of 2740 cpm corrected for background. A 5 ml sample of the patient's plasma provided 3021 cpm corrected for background. Compute the patient's plasma volume.

2. The following count rates corrected for background, were recorded 24 hr after administration of 10 μCi of Na^{131}I to a patient:

patient's thyroid	1421 cpm
patient's thyroid shielded by lead filter	281 cpm
standard in neck phantom	4467 cpm
standard in neck phantom shielded by lead filter	187 cpm

What was the percent uptake?

3. Describe a procedure in which a radioactive nuclide may be used to estimate the volume of a large container.

4. A PBI conversion ratio test yielded the following results:

count rate for 5 ml of plasma	1252 cpm
count rate for PBI in 5 ml of plasma	697 cpm
count rate for background	31 cpm

Determine and evaluate the PBI conversion ratio.

REFERENCES

[1] Powsner, E., and D. Raeside. 1971. Diagnostic nuclear medicine. Grune and Stratton, New York. p. 339.

[2] Hendee, W. 1970. Medical radiation physics. Year Book Medical Publishers, Chicago. p. 412.

[3] Hamolsky, M., M. Stein, and A. Freedberg. 1957. The thyroid hormone-plasma protein complex in man. II. A new *in vitro* method for study of "uptake" of labeled hormonal components by human erythrocytes, J. Clin. Endocrin. Metabl. 17:33.

[4] Chase, G., and J. Rabinowitz. 1967. Principles of radioisotope methodology, 3rd ed. Burgess Publishing Co., Minneapolis. p. 547.

21

Scanning and Whole-Body Counting

Some elements in ionic form are concentrated in specific tissues or regions of the body. If γ-emitting radioactive isotopes of these elements are administered orally or intravenously, then usually their distribution and concentration will be identical to those for the stable counterparts of the radioactive isotopes. For example, isotopes of iodine (^{123}I, ^{125}I, ^{131}I, and ^{132}I) are collected by the thyroid gland after administration as NaI; ^{85}Sr and ^{87m}Sr administered as $Sr(NO_3)_2$ are concentrated in bone; and ^{67}Ga injected as gallium citrate is distributed in bone, liver, and, frequently, soft tissue tumors. Other nuclides can be concentrated in certain tissues by incorporating the nuclides into appropriate compounds. For example, ^{99m}Tc is collected by the liver when injected intravenously as colloidal sulfide, ^{131}I as macroaggregated serum albumin is deposited relatively uniformly throughout the capillaries of the lungs, and radioactive isotopes of mercury (^{197}Hg and ^{203}Hg) are distributed to the brain and kidneys when injected as labeled chlormerodrin. Some of the nuclides and compounds used commonly as imaging agents in clinical medicine are listed in Table 21-1.

After a radioactive nuclide has concentrated in a particular organ or region of the body, its distribution within the organ or region may be determined with an imaging device such as a rectilinear scanner or gamma camera. Areas of increased or reduced activity in the image of the nuclide distribution may reflect an abnormal condition such as nonfunctional tissue or the presence of a tumor, an abscess, a cyst, etc.

272

TABLE 21-1 Selected Scanning Agents

Organ	Nuclide	Agent	Physical half-life, days	Administered activity, μCi	Time of scanning after administration, hr
Brain	^{99m}Tc	Pertechnetate	0.25	5,000–10,000	0.2–1.0
	^{113m}In	DTPA^a	0.07	10,000–20,000	0.2–1.0
	^{131}I	Serum albumin	8	300–500	0.1 – 1.0
	^{197}Hg	Chlormerodrin	2.7	700–1,500	2–6
	^{203}Hg	Chlormerodrin	47	500–700	2–4
Bone	^{18}F	Fluoride	0.08	1,000–1,500	1.0–1.5
	^{85}Sr	Strontium nitrate	65	50–100	48–72
	^{87m}Sr	Strontium nitrate	0.12	1,000–3,000	1–4
Kidney	^{99m}Tc	DTPA^a	0.25	2,000–5,000	1–2
	^{131}I	Hippuran	8	30–100	Immediate
	^{197}Hg	Chlormerodrin	2.7	200–400	1–5
	^{203}Hg	Chlormerodrin	47	100–150	1–5
Liver	^{99m}Tc	Technetium sulfide (colloidal)	0.25	1,000–3,000	0.2–1.0
	^{113m}In	Fe-^{113m}In (colloidal)	0.07	1,000–1,500	0.1–0.2
	^{131}I	Rose Bengal	8	100–200	0.2–0.5
	^{131}I	Microaggregated albumin	8	300–500	0.2–0.8
	^{198}Au	Colloidal gold	2.70	100–300	0.5–4.0
Lung	^{51}Cr	Macroaggregated serum albumin	27.8	800–1,000	0.1–2.0
	^{99m}Tc	Albumin microspheres	0.25	1,000–3,000	Immediate
	^{99m}Tc	Macroaggregated serum albumin	0.25	1,000–3,000	Immediate
	^{113m}In	Iron hydroxide	0.07	2,000–3,000	Immediate
	^{131}I	Macroaggregated serum albumin	8	250–350	0.1–2.0
Pancreas	^{75}Se	Selenomethionine	120	200–300	0.2–0.5
Placenta	^{51}Cr	Erythrocytes	27.8	30–150	0.1–0.2
	^{99m}Tc	Serum albumin	0.25	500–1,000	0.1–1.0
	^{113m}In	Transferrin	0.07	1,000–3,000	0.1–0.2
	^{131}I	Serum albumin	8	1–5	0.1–1.0

^{a}DTPA = diethylenetriaminepentaacetic acid.

273

TABLE 21-1 (Cont.)

Organ	Nuclide	Agent	Physical half-life, days	Administered activity, μCi	Time of scanning after administration, hr
Pericardial effusion	99mTc	Pertechnetate	0.25	10,000	Immediate–0.5
	99mTc	Serum albumin	0.25	2,000	0.1–0.5
	131I	Cholografin	8	250–500	0.1–0.5
Spleen	51Cr	Damaged erythrocytes	27.8	300–1,000	1–24
	99mTc	Technetium sulfide (colloidal)	0.25	1,000–3,000	0.2–0.4
	197Hg	MHP[b]	2.7	300–500	0.5–2.0
Thyroid	99mTc	Pertechnetate	0.25	1,000–2,000	0.5–1.0
	125I	Sodium iodide	60	50–100	24–48
	131I	Sodium iodide	8	30–80	24–48

[b]MHP = mercurihydroxypropane.

Sequential images sometimes furnish additional information about the functioning of a particular organ.

Rectilinear Scanners

A rectilinear scanner consists of a radiation detector which moves in a plane parallel to the patient and in synchrony with a recording device which moves across an imaging medium such as photographic film. In most scanners, the detector is a NaI detector 3–5 in. in diameter. Between the detector and recording device is a pulse-height analyzer which transmits voltage pulses of a selected size to the recording device (Figure 21-1). A dual probe rectilinear scanner is shown in Figure 21-2.

To restrict the field of view of the moving detector, a collimator may be attached to the detector. The collimator transmits to the detector only those γ-ray photons which originate within a limited region of the patient. Various collimators are illustrated in Figure 21-3. Multihole-focused collimators are used for most scanning procedures.

The focal distance of a focused collimator is the distance between the face of the collimator and the location where rays converge after passing through holes in the collimator. Isoresponse curves depict the change in detector response as a point source of radiation is moved parallel and perpendicular to the collimator face (Figure 21-4). Usually, isoresponse

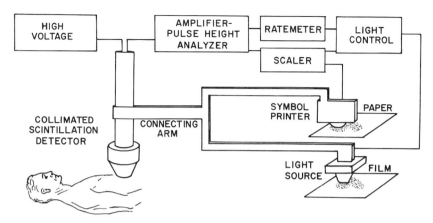

Figure 21-1 Diagram of a rectilinear scanner.

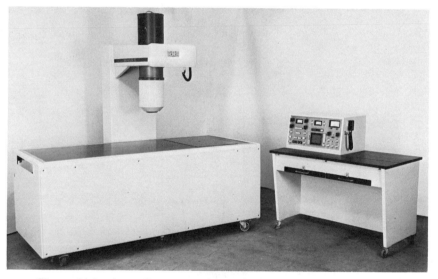

Figure 21-2 A rectilinear scanner with 5 in. diameter NaI detectors above and below the table. (Courtesy of Ohio Nuclear, Inc.)

curves are described as a percent of the response obtained with the source on the collimator axis and at the focal distance below the collimator face.

The resolution distance for a detector-collimator combination describes the capability of the combination to resolve adjacent areas of activity in the patient. The resolution distance may be defined as the

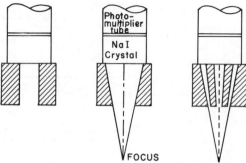

Figure 21-3 Collimators used with rectilinear scanners: left, single bore cylindrical collimator; center, single bore tapered collimator; right, multihole focused collimator.

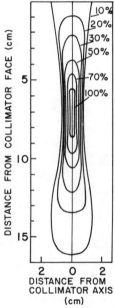

Figure 21-4 Isoresponse curves measured with a ^{133}Ba point source and a Picker #2102 37 hole focused collimator.

diameter of the 50% isoresponse curve in a plane parallel to the collimator face and at the focal distance below the face. Resolution improves as the resolution distance decreases. For a particular detector, the resolution distance may be reduced by choosing a collimator with holes of smaller diameter or by using a longer collimator or one with a shorter focal distance.

To obtain a photographic image of the distribution of activity in a selected anatomic region, the count rate from the detector is used to modulate the brightness or pulse duration of a light source (for example, a small cathode-ray tube) which moves across a photographic film in synchrony with the moving detector. Fluctuations in activity across the anatomic region are displayed as variations in optical density across the processed film. The distribution of activity may be imaged also in a paper dot scan. With this display technique, fluctuations in activity are represented as variations in the density of symbols printed on paper. The symbol printer prints a mark on the paper each time a selected number of pulses has been accumulated by a scaling circuit. Alternately, a conducting stylus may be used to burn a small dot in heat-sensitive Teledeltos paper whenever a selected number of pulses has been accumulated. With either of the methods, a region of increased activity in the patient is represented in the dot scan as an area of increased symbol density.

Positron scanning is a technique for achieving high spatial resolution of the distribution of radioactivity in a particular anatomic region. This technique involves the detection of 511 keV photons produced as positrons annihilate after their emission by a nuclide within the patient. The technique requires two moving NaI detectors, one on each side of the region to be scanned. Some of the annihilation photons emerging from the region reach both detectors simultaneously. Signals from the detectors are directed to a coincidence circuit; this circuit transmits a pulse to the display device only if it receives pulses simultaneously from both detectors. The display device indicates the highest count rate when the region of greatest activity is centered along a line through both detectors. If the region of greatest activity is nearer one detector, then the pulse rate also is greater from this detector. Pulse rate inequalities between the two detectors may be analyzed by a difference scaler to determine the position of greatest activity along a line between the detectors.

A variety of modifications of rectilinear scanning techniques have been developed to achieve improved spatial resolution and more rapid collection of data. These modifications are discussed in various texts and review articles [1-3].

Stationary-Imaging Devices

In a stationary-imaging device, the detector is exposed simultaneously to γ-ray photons from the entire region for which an image of the distribution of radioactivity is desired. Consequently, the image is formed more rapidly with a stationary-imaging device than with a rectilinear scanner, because the field of view of a scanner is restricted at each moment to only a small part of the region to be imaged. In studies of the functional capability of a particular organ, sequential images separated by short intervals of time may be obtained with a stationary-imaging device. Many of these instruments utilize a digital computer to enhance the collection and manipulation of data.

The stationary-imaging device employed most commonly is the single crystal γ camera developed by Anger in 1956[4]. The γ camera utilizes a 0.5 in. thick NaI crystal coupled optically to an array of 19 or more photomultiplier tubes (Fig. 21-5). Crystals vary in diameter from 11.5 to 13.5 in., and are large enough to encompass most organs in the body. When a γ-ray photon interacts in the crystal, light diffuses from the site of interaction toward the photocathodes of the photomultiplier tubes. The amount of light reaching a particular photocathode and, consequently, the size of the signal from the photomultiplier tube, decreases with increasing distance between the photocathode and the site of interaction.

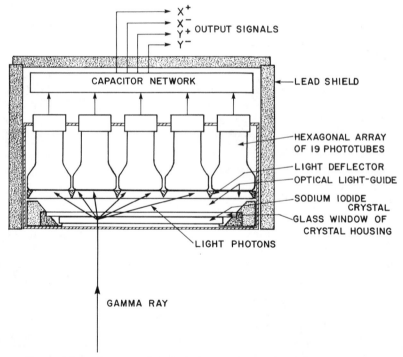

Figure 21-5 Detector for a single crystal γ camera. [From H. Anger. 1968. Sensitivity and resolution of the scintillation camera. *In* A. Gottschalk and R. Beck (ed.), Fundamental problems in scanning. Charles C. Thomas, Springfield, Ill. p. 118.]

The signals from the photomultiplier tubes are transmitted to logic circuits, where their relative amplitudes are analyzed to determine the position of a light spot on a cathode-ray screen which corresponds to the site of interaction of the γ-ray in the crystal. If the sum of all incoming signals corresponds to complete absorption of the γ-ray in the crystal, then a short burst of electrons is allowed to flow across the cathode-ray tube and produce a flash of light at the desired position on the screen. A camera may be attached to the screen to photograph the light flashes; usually, 50,000–500,000 flashes are required to produce an acceptable image of the distribution of radioactivity.

Three types of collimators are used with a γ camera (Figure 21-6). A pinhole collimator furnishes a field of view which increases with the distance between the collimator and the region to be studied, and is used primarily for studies of very small organs (such as the thyroid gland) and, occasionally, of large regions (an enlarged liver or the lungs) which cannot be encompassed by a parallel-multihole collimator. The resolution distance increases with increasing separation of the collimator and the region to be examined.

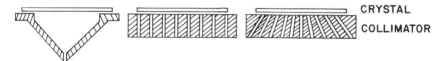

CRYSTAL

COLLIMATOR

Figure 21-6 Collimators used with a γ camera. Left, pinhole; center, parallel multihole; right, diverging.

Parallel-multihole collimators consist of an absorbing plate, usually tungsten alloy, with hundreds of small parallel holes. Although the face of the collimator is usually positioned as close as possible to the region to be imaged, the image does not change appreciably as the collimator and detector are moved away from the region.

Occasionally, the field of view of a parallel-multihole collimator is too small to encompass large regions such as the lungs. To obtain a complete image of large regions, a diverging collimator may be used. This collimator is a multihole collimator containing holes of uniform diameter which diverge radially in a direction from the crystal toward the patient. This collimator provides an expanded field of view with only a small loss in detector sensitivity.

Multiple interactions of a single photon in the crystal of a γ camera reduce the resolution (that is, increase the resolution distance), because the light flash is positioned on the screen of the cathode-ray tube at a location intermediate among those which correspond to the interaction sites.

This reduction in resolution is more severe for higher energy photons, because these photons are absorbed more often by multiple interactions. Resolution loss is caused also by errors in the positioning of the light flash resulting from fluctuations in the number of electrons liberated in a photomultiplier tube for a given number of light photons received by the photocathode. This contribution to resolution loss increases rapidly with decreasing photon energy, and establishes a lower limit of about 70 keV for acceptable resolution in a γ camera. An upper limit of about 700 keV is established by the excessive penetration of collimator septa by photons of higher energy. For γ cameras and collimators currently available, the smallest resolution distance obtainable is about 1 cm, a value comparable to that for a rectilinear scanner (Figure 21-7)[5].

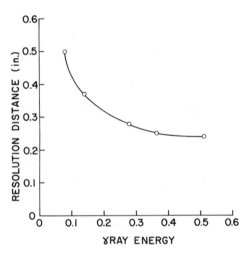

Figure 21-7 Resolution distance of a 11 × 1/2 in. NaI detector for a γ camera. Data were measured with an analyzer window 1.5 times the full width at half-maximum. [From H. Anger. 1968. Sensitivity and resolution of the scintillation camera. *In* A. Gottschalk and R. Beck (ed.), Fundamental problems in scanning Charles C. Thomas, Springfield, Ill. p. 124.]

For dynamic function studies, sequential images displayed on a cathode-ray screen may be recorded on motion picture film. Alternately, the x and y coordinates of each interaction in the crystal may be recorded on video tape. The tape may be played back at a desired speed to furnish an image on a television monitor.

Whole-Body Counters

Whole-body counters are used to identify and quantify radioactive nuclides which may be present in the body of humans or animals. Data are obtained by positioning detectors at strategic locations outside the body to detect γ-ray photons, annihilation radiation, or bremsstrahlung escaping from the body. For accurate measurements, the detectors must be positioned to minimize variations in response caused by changes in the distribution of radioactivity in the body. This requirement may be satisfied in two ways:

(1) One large annular detector may be chosen which surrounds the body almost completely. This type of detector provides essentially 4π counting geometry and intercepts almost all photons which leave the body.

(2) One or more small detectors may be positioned at a chosen location near the body or moved along a track parallel to the long axis of the body. With this arrangement, the number of photons intercepted by the detectors is a representative fraction of the photons emitted from the body during the period of measurement.

Although a few counters with high-pressure ion chambers or other gas-filled detectors are still in use, inorganic crystals (usually NaI), plastic solids, and liquid scintillators usually are preferred for whole-body counters. For a single or multiple detector arrangement of stationary solid detectors designed to detect photons from the human body, the patient usually is positioned in either a supine or a sitting posture (Figure 21-8). For an arrangement of moving detectors, the patient usually is

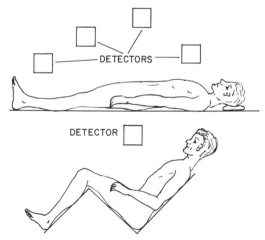

Figure 21-8 Geometries for whole-body counting.

supine. To reduce the background count rate, massive shielding of steel or lead may be added to the walls, ceiling and floor of the counting facility. Alternately, a shadow shield may be employed. A shadow shield consists of a collimating shield around the detector and a large shield behind the patient which is shaped to project a shadow over the area viewed by the detector (Figure 21-9).

A liquid scintillation whole-body counter usually consists of a long cylindrical tank containing liquid scintillation cocktail between the inner and outer walls, and a number of photomultiplier tubes coupled to the

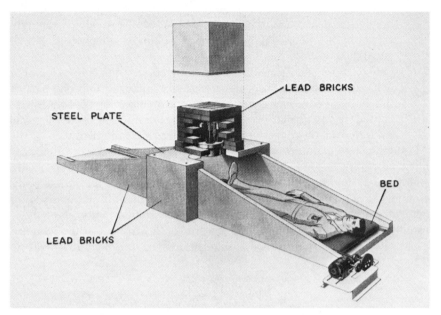

Figure 21-9 Moving single detector whole-body counter with a shadow shield. [From H. Palmer and W. Roesch. 1965. A shadow shield whole-body counter. Health Phys. 11:1216.]

outer wall. The supine patient is counted while positioned inside the cylindrical tank. A pulse-height spectrum obtained with a typical whole-body counter designed for humans is shown in Figure 21-10.

Although radionuclides in small animals may be identified with a whole-body counter designed for humans, a few counters are available commercially for identification and measurement of radioactivity in animals. The small animal counter shown in Figure 21-11 uses a liquid scintillation cocktail as a detector in a tank about 12 in. in outside diameter and 12 in. long. The animal (or sample) is positioned within a 4 in.

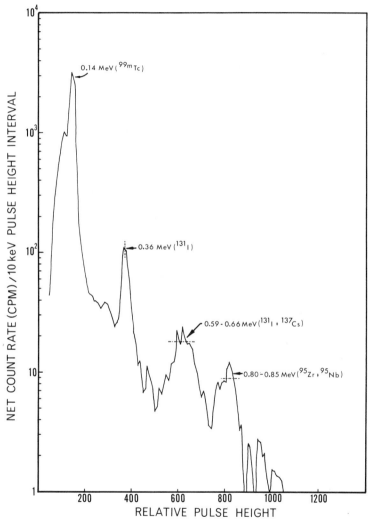

Figure 21-10 Pulse-height spectrum furnished by a whole-body counter. (Courtesy of Colorado Department of Health.)

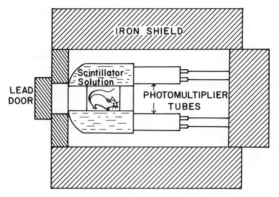

Figure 21-11   Detector and shield of a typical small animal counter.

opening in the center of the tank. The scintillation cocktail is monitored by seven photomultiplier tubes mounted on a flat plate at one end of the tank. Steel plates 7 in. thick are used as a background shield.

REFERENCES

[1] Hendee, W. 1970. Medical radiation physics. Year Book Medical Publishers, Chicago. p. 410.

[2] Mallard, J. 1966. Medical radioisotope visualization: A review of "scanning." Internat. J. Appl. Radiat. Isotopes 17:205.

[3] Gottschalk, A., and R. Beck (ed). 1968. Fundamental problems in scanning. Charles C. Thomas, Springfield, Ill.

[4] Anger, H. 1958. Scintillation camera. Rev. Scient. Inst. 29:27.

[5] Anger, H. 1964. Scintillation camera with multichannel collimator. J. Nuclear Med. 5:515.

22

Radiation Exposure and Dose

The effects of ionizing radiation on a particular chemical process or biologic specimen often are explained in terms of the amount of radiation to which the process or specimen has been exposed, or the amount of energy which the process or specimen has absorbed. The radiation exposure in units of roentgens is a reflection of the amount of ionization which the radiation would produce in a small mass of air. The absorbed dose in units of rads is a measure of the energy absorbed from the radiation per unit mass of irradiated material. For x- and γ-ray photons, an exposure in roentgens and the corresponding absorbed dose in rads are related by the f-factor, a coefficient which varies with the energy of the photons and the nature of the irradiated material. The dose equivalent in units of rems and the relative biologically effective (RBE) dose in rems are modifications of the absorbed dose which include consideration of the effectiveness of the radiation for producing observable effects.

Radiation Exposure

Ion pairs (electrons and positive ions) are produced as ionizing radiation interacts with atoms of an attenuating medium. These ion pairs dissipate their energy by ionizing nearby atoms, creating additional ion pairs. The total number of ion pairs produced varies with the energy deposited in the medium by the radiation. If the attenuating medium is air, and if Q is the total charge (positive or negative) liberated as x- or γ-ray photons interact

285

in a small volume of air of mass m, then the radiation exposure X at the location of the small volume is

$$X = \frac{Q}{m}$$

where Q includes ion pairs produced both inside and outside of the small volume of air. The unit of radiation exposure is the roentgen (R).

$$R = 2.58 \times 10^{-4} \text{ coul/kg of air}$$

The roentgen is used only for x- and γ-ray photons with energy less than about 3 MeV. It is not used for x- and γ-ray photons above about 3 MeV, or for charged particles or neutrons.

Since the W quantity is 33.7 eV/IP for x- and γ-rays in air, the energy absorbed in a unit mass of air during an exposure X in roentgens is

energy absorbed in air

$$= \frac{X(\text{R})(2.58 \times 10^{-4} \text{ coul/kg-R})(33.7 \text{ eV/IP})(1.6 \times 10^{-19} \text{ J/eV})}{1.6 \times 10^{-19} \text{ coul/IP}}$$
$$= 86.9 \times 10^{-4}(\text{J/kg-R})X(\text{R})$$

It is not possible to measure all of the ion pairs produced as energy is deposited in a small volume of air exposed to x- or γ-radiation. For example, ion pairs may not be collected if they are produced by electrons escaping from the small air volume. However, the volume of air may be chosen so that energy expended outside the air volume by ion pairs created within is compensated exactly by energy lost inside the small volume of air by ion pairs which originate outside. This condition, referred to as *electron equilibrium*, must be achieved before meaningful measurements of radiation exposure can be attained. Ionization chambers used to calibrate exposure rates for beams of x- and γ-ray photons are constructed with air-equivalent walls which provide electron equilibrium[1].

Absorbed Dose

The response of a particular material to radiation depends upon the energy deposited in the material by the radiation. The energy deposited in a medium by any type of ionizing radiation may be described in units of rads. The rad (an acronym for radiation absorbed dose) is a unit of absorbed dose, and represents the absorption of 10^{-2} J of energy per kilogram (or 10^2 ergs of energy per gram) of absorbing material.

$$1 \text{ rad} = 10^{-2} \text{ J/kg} = 100 \text{ erg/g}$$

The absorbed dose D in rads delivered to a small mass m in kilograms is:

$$D = \frac{E/m}{10^{-2}\,\text{J/kg-rad}}$$

where E is the total energy in joules deposited in the mass, corrected for energy removed from the medium in any way (for example, bremsstrahlung, characteristic radiation, etc.).

Example 22-1

A dose of 5000 rad is delivered uniformly to a 100 g mass of tissue. How much energy in joules is absorbed by each gram of tissue and by the entire tissue mass?

$$D = \frac{E/m}{10^{-2}\,\text{J/kg-rad}}$$

$$
\begin{aligned}
E/m &= (10^{-2}\,\text{J/kg-rad})D \\
&= (10^{-2}\,\text{J/kg-rad})(5000\,\text{rad})(10^{-3}\,\text{kg/g}) \\
&= 0.05\,\text{J/g} \\
E &= (0.05\,\text{J/g})(m) \\
&= (0.05\,\text{J/g})(100\,\text{g}) \\
&= 5\,\text{J}
\end{aligned}
$$

Consequently, each gram of tissue receives 0.05 J, and the total mass of tissue receives 5 J.

f-Factor

The energy absorbed in air during an exposure X in roentgens has been noted above to be

$$\text{energy absorbed in air} = (86.9 \times 10^{-4}\,\text{J/kg-R})X(\text{R})$$

Since 1 rad $= 10^{-2}\,\text{J/kg}$, an exposure of 1 R provides an absorbed dose of 0.869 rads in air, and the expression above may be stated

$$D_{\text{air}}\,(\text{rad}) = 0.869X(\text{R})$$

where D_{air} is the absorbed dose in air corresponding to an exposure X. For any medium, the dose D_{med} in rads corresponding to an exposure X is

$$D_{\text{med}} = D_{\text{air}}\frac{[(\mu_{en})_m]_{\text{med}}}{[(\mu_{en})_m]_{\text{air}}} = 0.869\frac{[(\mu_{en})_m]_{\text{med}}}{[(\mu_{en})_m]_{\text{air}}}X$$

where $[(\mu_{en})_m]_{\text{med}}$ and $[(\mu_{en})_m]_{\text{air}}$ are the energy absorption coefficients of the medium and air, respectively, for the x- or γ-ray photons of interest. The product of 0.869 and $[(\mu_{en})_m]_{\text{med}}/[(\mu_{en})_m]_{\text{air}}$ is termed the *f*-factor, the conversion factor from roentgens to rads for an irradiated medium

$$f = 0.869 \frac{[(\mu_{en})_m]_{med}}{[(\mu_{en})_m]_{air}}$$

Hence

$$D_{med} = fX$$

Values for the f-factor in soft tissue and bone, averaged over typical x-ray spectra, are presented in Figure 22-1 as a function of the half-value layer of the x-ray beam.*

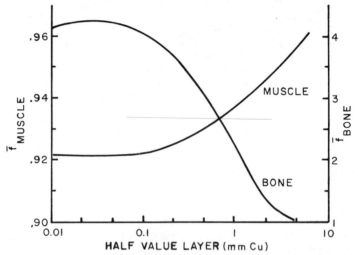

Figure 22-1 Mean conversion factor $\bar{f}$ from roentgens to rads in muscle and compact bone for x-ray beams with different half-value layers.

Relative Biologically Effective Dose and Dose Equivalent

Often, the effects of irradiation depend not only upon the amount of energy absorbed in an irradiated medium, but also upon the distribution of the absorbed energy within the medium. For equal absorbed doses, different types of ionizing radiation may vary in their efficiency to elicit a particular response. For biologic specimens, the RBE describes the effectiveness or efficiency with which a particular type of radiation evokes a certain biologic effect. The RBE is determined by comparing results

*The half-value layer for a beam of x- or γ-ray photons is the amount of attenuating material necessary to reduce the exposure rate of the beam to half. Sometimes the word intensity is substituted for exposure rate in this definition.

obtained with the radiation in question to those observed with a reference radiation (for example, medium energy x-rays or ^{60}Co γ-rays).

RBE

$$= \frac{\text{dose of reference radiation required to produce a particular response}}{\text{dose of radiation in question to produce the same response}}$$

For a particular type of radiation, the RBE may differ from one biologic response to another. The RBE dose in rems (acronym for roentgen equivalent man) is the product of the absorbed dose D in rads and the appropriate value of RBE

$$\text{RBE dose (rem)} = D \text{ (rad)} \times \text{RBE}$$

It has been suggested that the concept of RBE dose should be limited to descriptions of radiation dose in radiation biology[2].

Often, the effectiveness with which different types of radiation produce a particular biologic effect varies with the LET of the radiation. The dose equivalent (DE) in rems is the product of the absorbed dose in rads and a quality factor (QF) which varies with the LET of the radiation.

$$\text{DE (rem)} = D \text{ (rad)} \times \text{QF}$$

The dose equivalent reflects a recognition of differences in the effectiveness of different radiations to inflict overall biologic damage, and is used primarily during computations associated with radiation protection. Quality factors are listed in Table 22-1 as a function of the LET and in Table 22-2 for different types of radiation. These quality factors should be used for determination of shielding requirements and for computation of radiation doses for personnel working with or near sources of ionizing radiation.

TABLE 22-1 *Relation Between Linear Energy Transfer and Quality Factor*[a]

Average LET in water, keV/μ	Quality factor
3.5 or less	1
7	2
23	5
53	10
175	20

[a]From recommendations of the International Commission on Radiological Protection[3].

TABLE 22-2 *Quality Factors for Different Radiations*[a]

Type of radiation	Quality factor
x-rays, γ-rays, electrons or positrons	1
Neutrons, energy < 10 keV	3
Neutrons, energy > 10 keV	10
Protons	1–10
α Particles	1–20
Fission fragments, recoil nuclei	20

[a]These data should be used only for purposes of radiation protection[3].

Example 22-2

A person receives a whole-body dose of 200 mrad, of which 150 mrad is attributed to ^{60}Co γ-radiation and 50 mrad is attributed to neutrons with energy above 10 keV. What is the dose equivalent DE for the person in mrems?

$$DE \text{ (mrem)} = [D \text{ (mrad)} \times QF]_{\gamma\text{-rays}} + [D \text{ (mrad)} \times QF]_{neutrons}$$
$$= (150 \text{ mrad})(1) + (50 \text{ mrad})(10)$$
$$= 650 \text{ mrem}$$

Example 22-3

A person accidentally ingests a small sample of tritium. The average dose to the gastrointestinal tract is estimated to be 800 mrads. What is the dose equivalent in mrems?

$$DE \text{ (mrem)} = D \text{ (mrad)} \times QF$$
$$= (800 \text{ mrad})(1)$$
$$= 800 \text{ mrem}$$

In the expression for dose equivalent, a distribution factor (DF) may be included to compensate for changes in the radiation response caused by nonuniform distribution of a radioactive nuclide in the body. Other terms also may be added to describe the influence of additional factors upon the radiation response. Then the expression for dose equivalent is written

$$DE \text{ (rem)} = D \text{ (rad)} \times QF \times DF \ldots$$

Exposure Rate Constant*

The exposure rate at 1 m from a point source of a radioactive nuclide of 1 Ci activity is known as the exposure rate constant Γ_∞ for the nuclide.

*The exposure rate constant Γ_∞ replaces, but is not identical to, the specific γ-ray constant Γ[4].

With this constant, the exposure rate at a prescribed distance from a radio-active source may be estimated (Examples 22-4 and 22-5). The exposure rate constant in units of $(R-m^2)/(hr-Ci)$ for a particular nuclide may be computed with the expression

$$\Gamma_\infty = 19.9 \sum n_i E_i (\mu_{en})_i$$

where n_i is the fractional emission of x- or γ-ray photons of energy E_i in MeV, and $(\mu_{en})_i$ is the mass energy absorption coefficient cm^2/g of air for photons of energy E_i. Values of Γ_∞ for a few nuclides are included in Table 23-2.

Example 22-4
Estimate the exposure rate at 50 cm from a small source of ^{60}Co of 100 mCi activity. The exposure rate constant is 1.29 R-m²/hr-Ci for ^{60}Co.

$$\begin{aligned}
\text{Exposure rate} &= \Gamma_\infty A/d^2 \\
&= \frac{(1.29 \text{ R-m}^2/\text{hr-Ci})(0.1 \text{ Ci})}{(0.5 \text{ m})^2} \\
&= 0.516 \text{ R/hr} \\
&= 516 \text{ mR/hr}
\end{aligned}$$

Example 22-5
In 95% of the decays of ^{137}Cs, a β particle of $E_{max} = 0.51$ MeV is followed by an isomeric transition of 0.662 MeV. Of the isomeric transitions, 90% are achieved by emission of a γ-ray of 0.662 MeV, and 10% are achieved by internal conversion. The mass energy absorption coefficient of air is 0.029 cm²/g for photons of 0.662 MeV. Estimate the exposure rate constant Γ_∞ for ^{137}Cs.

The fractional emission n for γ-rays from ^{137}Cs is

$$n = (0.95)(0.90) = 0.85$$

The x-rays accompanying internal conversion are emitted with the following energies, fractional emissions, and mass energy absorption coefficients in air.

Energy, MeV	Fractional emission	Mass energy absorption coefficient, cm²/g
0.032	0.056	0.132
0.036	0.011	0.099
0.037	0.002	0.091

$$\Gamma_\infty = 19.9\Sigma n_i E_i (\mu_{en})_i$$
$$= 19.9[(0.85)(0.66)(0.029) + (0.056)(0.032)(0.132)$$
$$+ (0.011)(0.036)(0.099) + (0.002)(0.037)(0.091)]$$
$$= 19.9[(0.0165) + (0.0002) + (0.0000) + (0.0000)]$$
$$= 0.333 \text{ R-m}^2/\text{hr-Ci}$$

PROBLEMS

1. How many ion pairs are produced in 1 cm³ of air during an exposure of 50 R?

2. What exposure in roentgens will result in the absorption of 1000 erg/g of air?

3. An individual ingests some radioactive material which delivers to the lining of the GI tract 1800 mrad due to α particles (QF = 20) and 1100 mrad due to γ-rays. What is the dose equivalent in mrems?

4. A ^{60}Co source provides an exposure rate of 92 R/min at a distance of 100 cm. A rat 7.5 cm thick is positioned in the beam with the nearest surface 100 cm from the source and irradiated for 5 min. What is the absorbed dose in rads delivered to soft tissue at the midline of the rat? The fractional dose is 0.82 at the midline compared to the surface, and the f-factor is 0.96 for ^{60}Co γ-rays in soft tissue.

5. The exposure rate constant is approximately 0.22 R-m²/hr-Ci for ^{131}I. Estimate the exposure rate at a distance of 50 cm from a small sample of ^{131}I of 200 mCi activity.

6. From the following data, estimate the exposure rate constant for ^{125}I.

Radiation	Energy, MeV	Fractional emission	Energy absorption coefficient, cm²/g
γ	0.0355	0.068	0.103
x-ray	0.0275	0.738	0.239
x-ray	0.0272	0.378	0.250
x-ray	0.0310	0.199	0.140
x-ray	0.0318	0.0413	0.133

REFERENCES

[1] Hendee, W. 1970. Medical radiation physics. Year Book Medical Publishers, Chicago. Chap. 8.

[2] International Commission on Radiation Units and Measurements. 1968. Radiation quantities and units. ICRU Report 11. Government Printing Office, Washington, D.C.

[3] International Commission on Radiological Protection. 1966. Recommendations of the international commission on radiological protection. ICRP Publ. 9, Pergamon Press, Oxford.

[4] International Commission on Radiation Units and Measurements. 1971. Radiation quantities and units. ICRU Report 19. Government Printing Office, Washington, D.C.

23

Internal Absorbed Dose

Radioactive nuclides frequently are administered orally or intravenously to a patient for diagnosis or treatment of a particular disorder or for investigation of the pathways of a particular compound in the body. Occasionally, a radioactive nuclide is inhaled, ingested, or absorbed accidentally by an individual working with or near radioactive materials. Whenever radioactive nuclides are deposited internally, the absorbed dose to various body organs should be estimated.

Standard Man

To compute the absorbed dose delivered to an organ or to the whole body by radioactivity inside the body, the mass of the organ or whole body must be estimated. The International Commission on Radiological Protection (ICRP) has developed a "standard man" which provides estimates of the average mass of organs in the adult[1]. These estimates are included in Table 23-1.

Effective Half-Lives

The amount of a radioactive nuclide in a particular organ changes with time after intake of the nuclide into the body. The change reflects not only radioactive decay of the nuclide, but also the influence of physiologic processes which move chemical substances into and out of various organs

294

TABLE 23-1 *Average Mass of Organs in Adult Human Body[a]*

Organ	Mass, g	Total Body,[b] %
Total body[b]	70,000	100
Muscle	30,000	43
Skin and subcutaneous tissue[c]	6,100	8.7
Fat	10,000	14
Skeleton		
without bone marrow	7,000	10
red marrow	1,500	2.1
yellow marrow	1,500	2.1
Blood	5,400	7.7
Gastrointestinal tract[b]	2,000	2.9
Contents of GI tract		
lower large intestine	150	
stomach	250	
small intestine	1,100	
upper large intestine	135	
Liver	1,700	2.4
Brain	1,500	2.1
Lungs (2)	1,000	1.4
Lymphoid tissue	700	1.0
Kidneys (2)	300	0.43
Heart	300	0.43
Spleen	150	0.21
Urinary bladder	150	0.21
Pancreas	70	0.10
Salivary glands (6)	50	0.071
Testes (2)	40	0.057
Spinal cord	30	0.043
Eyes (2)	30	0.043
Thyroid gland	20	0.029
Teeth	20	0.029
Prostate gland	20	0.029
Adrenal glands or suprarenal (2)	20	0.029
Thymus	10	0.014
Ovaries (2)	8	0.011
Hypophysis (Pituitary)	0.6	8.6×10^{-6}
Pineal gland	0.2	2.9×10^{-6}
Parathyroids (4)	0.15	2.1×10^{-6}
Miscellaneous (blood vessels, cartilage, nerves, etc.)	390	0.56

[a]From International Commission on Radiological Protection. 1969. Report of Committee II: Permissible dose for internal radiation. Pergamon Press, New York.
[b]Does not include contents of the gastrointestinal tract.
[c]The mass of the skin alone is about 2000 g.

in the body. The effective half-life for uptake or elimination of a nuclide in a particular organ is the time required for the activity of the nuclide to increase or decrease to half its maximum activity in the organ. The effective half-life for uptake (T_{up}) is computed from the half-life T_1 for physiologic uptake, excluding radioactive decay, and the half-life $T_{1/2}$ for radioactive decay. The effective half-life for elimination (T_{eff}) of a radioactive nuclide from an organ is computed from the half-life T_b for physiologic elimination, excluding radioactive decay, and the half-life for radioactive decay. Expressions for the computation of T_{up} and T_{eff} are

$$T_{up} = \frac{T_1 T_{1/2}}{T_1 + T_{1/2}}$$

$$T_{eff} = \frac{T_b T_{1/2}}{T_b + T_{1/2}}$$

where T_1 is the half-life for physiologic uptake, excluding radioactive decay, T_b is the half-life for physiologic elimination, excluding radioactive decay, and $T_{1/2}$ is the half-life for radioactive decay.

Example 23-1
Chlormerodrin labeled with ^{197}Hg is removed from the kidney with a half-life for physiologic elimination of 5.6 hr (0.23 days). The half-life is 2.7 days for radioactive decay of ^{197}Hg. What is the effective half-life for elimination of ^{197}Hg-chlormerodrin from the kidney?

$$
\begin{aligned}
T_{eff} &= \frac{T_b T_{1/2}}{T_b + T_{1/2}} \\
&= \frac{(0.23 \text{ days})(2.7 \text{ days})}{(0.23 \text{ days}) + (2.7 \text{ days})} \\
&= 0.21 \text{ days}
\end{aligned}
$$

The effective half-life for elimination of a nuclide always is shorter than either the half-life for physiologic elimination or the half-life for radioactive decay.

Example 23-2
Iodine is absorbed into the thyroid with a half-life of 5 hr for physiologic uptake. The half-life for radioactive decay of ^{123}I is 13 hr. What is the effective half-life for uptake of ^{123}I into the thyroid?

$$
\begin{aligned}
T_{up} &= \frac{T_1 T_{1/2}}{T_1 + T_{1/2}} \\
&= \frac{(5 \text{ hr})(13 \text{ hr})}{(5 \text{ hr}) + (13 \text{ hr})} \\
&= 3.6 \text{ hr}
\end{aligned}
$$

The effective half-life for uptake of a nuclide always is shorter than either the half-life for physiologic uptake or the half-life for radioactive decay.

Internal β Dose

The internal β dose (described more appropriately as the "dose due to nonpenetrating radiation") is the absorbed dose delivered to tissue by radiation which penetrates no farther than about 1 cm from its origin. Contributions to the internal β dose include:

(a) The absorbed dose delivered by negatrons and positrons emitted by radioactive nuclei
(b) The absorbed dose delivered by electrons released during internal conversion
(c) The absorbed dose delivered by Auger electrons
(d) The absorbed dose delivered by x-ray and γ-ray photons with energy less than 11.3 keV; characteristic x-rays which contribute to the internal β dose include:
  K-characteristic x-rays from elements with $Z < 35$
  L-characteristic x-rays from elements with $Z < 85$
  M-characteristic x-rays from all elements

An average β energy $\bar{E}_\beta$ may be computed for every radioactive nuclide, including those which decay without the emission of negatrons or positrons. For example, an average β energy may be determined for nuclides which decay solely by electron capture or isomeric transition. The average β energy is the average energy absorbed locally (within 1 cm of its origin) per disintegration of the nuclide. Included in Table 23-2 are average β energies for many radioactive nuclides used in biology and medicine.

Suppose that a homogeneous mass of tissue contains a uniform distribution of a particular radioactive nuclide at a concentration C in microcuries per gram. The mass of tissue is large enough to provide equilibrium between energy entering and leaving any region within the mass. The instantaneous dose rate R_β, expressed in rads per second, rads per hour, and rads per day, is

$$R_\beta \text{ (rad/sec)} = 5.92 \times 10^{-4} C\bar{E}_\beta$$
$$R_\beta \text{ (rad/hr)} = 2.13 C\bar{E}_\beta$$
$$R_\beta \text{ (rad/day)} = 51.2 C\bar{E}_\beta$$

where C is the concentration of the nuclide in microcuries per gram and $\bar{E}_\beta$ is the average β energy in million electron volts. It is apparent from

TABLE 23-2 Data for Selected Radioactive Nuclides Used in Biology and Medicine

Radioactive nuclide	Half-life	$\bar{E}_\beta$, MeV	Photon energies E_i, MeV	Fractional emission n_i	Δ_i, g-rad/μCi-hr	Γ_∞, R-m²/hr-Ci
³H	12.3 yr	0.0055	—	—	—	—
¹¹C	20.3 month	0.3942	0.511	1.996	2.1765	—
¹⁴C	5580 yr	0.0493	—	—	—	—
²²Na	2.6 yr	0.194	1.2746	1.00	2.7200	0.659
			0.511 (annihilation)	1.800	1.9629	
²⁴Na	15.0 hr	0.555	1.3685	0.999	2.9149	1.859
			2.7539	0.999	5.8599	
³²P	14.3 days	0.6948	—	—	—	—
³⁵S	87.2 days	0.0488	—	—	—	—
⁴²K	12.4 hr	1.4317	0.3100	0.00179	0.0012	0.139
			0.5240	0.18	0.5843	
⁴⁵Ca	165 days	0.0756	—	—	—	—
⁶⁰Co	5.26 yr	0.0949	1.1732	0.998	2.4939	1.29
			1.3325	1.00	2.8382	
⁶⁵Zn	245 days	0.0164	0.511 (annihilation)	0.0340	0.0400	0.296
			1.1150	0.4899	1.1657	
			0.2125	0.0080	0.0036	
			0.1061	0.0080	0.0018	
			0.0370	0.0089	0.0007	

these expressions that the instantaneous dose rate varies with changes in the concentration C of the nuclide in the mass of tissue. If the nuclide is eliminated exponentially, then

$$C = C_{\max}\, e^{-\ln 2(t/T_{\mathrm{eff}})}$$

where t is the time elapsed since the concentration was at its maximum value $C_{\max}$. If the assumption of exponential elimination is too great a simplification, then the elimination of the nuclide from the tissue must be described by a more complex expression[2]. If exponential elimination may be assumed, then the instantaneous dose rate in rads per day is

$$R_\beta\ (\text{rad/day}) = 51.2\bar{E}_\beta C_{\max}\, e^{-\ln 2(t/T_{\mathrm{eff}})}$$

For a finite interval of time which begins at $t = 0$ when $C = C_{\max}$, an expression for the cumulative internal β dose D_β is obtained by integrating the expression for R_β.* The result is

$$D_\beta\ (\text{rad}) = 73.8\bar{E}_\beta C_{\max} T_{\mathrm{eff}}\,(1 - e^{-\ln 2(t/T_{\mathrm{eff}})})$$

with T_{eff} and t expressed in days. The total absorbed dose delivered by energy absorbed locally as the nuclide is eliminated completely from the mass of tissue is obtained by letting t approach ∞ in the expression above. The exponential term approaches 0 and

$$D_\beta\ (\text{rad}) = 73.8\bar{E}_\beta C_{\max} T_{\mathrm{eff}}$$

If the effective half-life for uptake is very short compared to the effective half-life for elimination, then the expression for D_β above may be used to estimate the total β dose delivered by a radioactive nuclide from the moment it is distributed in a mass of tissue until the time it is eliminated completely. If the uptake of the nuclide must be considered, and if the uptake may be described exponentially, then the expression for the total cumulative β dose D_β may be written

$$D_\beta\ (\text{rad}) = 73.8\bar{E}_\beta C_{\max} T_{\mathrm{eff}}\left(1 - \frac{T_{\mathrm{up}}}{T_{\mathrm{eff}}}\right)$$

with T_{eff} and T_{up} expressed in days. If $T_{\mathrm{eff}} \geq 20(T_{\mathrm{up}})$, then the effective half-life for uptake (that is, the term $1 - (T_{\mathrm{up}}/T_{\mathrm{eff}})$) may be neglected with an error not greater than 5% for the total cumulative β dose.

Example 23-3
An injection of one mCi of ^{14}C is delivered intravenously to a 70 kg man in a form which will cause the radioactive material to remain in

*Derivations of the equations in this chapter are available in W. Hendee. 1970. Medical radiation physics. Year Book Medical Publishers, Chicago. Chap. 26.

the blood with a biologic half-life of 100 days. What are the instantaneous dose rate, the dose delivered over the first week, and the total dose to the blood?

From Table 23-1, the mass of blood is 5400 g in the standard 70 kg man. For ^{14}C, $\bar{E}_\beta = 0.05$ MeV, and the instantaneous dose rate is

$$R_\beta = 5.92 \times 10^{-4} C \bar{E}_\beta$$
$$= 5.92 \times 10^{-4} \left(\frac{1000\ \mu Ci}{5400\ g}\right)(0.05\ \text{MeV})$$
$$= 5.5 \times 10^{-6}\ \text{rad/sec}$$
$$= 5.5\ \mu\text{rad/sec}$$

Because the half-life for radioactive decay is very long (5580 years), the effective half-life equals the biologic half-life for elimination of ^{14}C from the blood. If this elimination is exponential, then the dose delivered over the first week is

$$D_\beta = 73.8 \bar{E}_\beta C_{max} T_{eff} (1 - e^{-\ln 2(t/T_{eff})})$$
$$= 73.8(0.05\ \text{MeV}) \left(\frac{1000\ \mu Ci}{5400\ g}\right)(100\ \text{days})(1 - e^{-\ln 2(7/100)})$$

$$= (0.0685)(1 - 0.952)$$
$$= 3.3\ \text{rad over the first week}$$

The total dose delivered to the blood during complete elimination of the radioactive material is

$$D_\beta = 73.8 \bar{E}_\beta C_{max} T_{eff}$$
$$= 73.8(0.05\ \text{MeV}) \left(\frac{1000\ \mu Ci}{5400\ g}\right)(100\ \text{days})$$
$$= 68\ \text{rad}$$

Internal γ Dose

The internal γ dose (described more appropriately as the "dose due to penetrating radiation") is the absorbed dose delivered to tissue by radiation which travels, on the average, 1 cm or more beyond its origin. This radiation includes γ-ray photons with energy above 11.3 keV, annihilation photons of 0.51 MeV, K-characteristic X-rays from elements with Z greater than 35, and L-characteristic X-rays from elements with Z greater than 85.

The size, shape, and location of a mass of tissue determines the ratio of the energy absorbed from penetrating radiation (that is, photons) traversing the mass of tissue to the total energy released as penetrating radiation during decay of radioactive nuclei in the same or a different mass of tissue. This ratio is termed the *absorbed fraction* ϕ.

$$\phi = \frac{\text{energy from penetrating radiation absorbed by a mass of tissue}}{\text{energy in the form of penetrating radiation emitted by nuclide in a mass of tissue}}$$

Absorbed fractions for spheres and ellipsoids containing uniform distributions of radioactivity are included in Tables 23-3 through 23-8. More complete tabulations of absorbed fractions are available in publications from the Society of Nuclear Medicine[3].

Suppose that a mass of tissue contains a uniform distribution of a particular photon-emitting nuclide at a concentration C in microcuries per gram. For photons of energy E_i, the instantaneous dose rate R_γ to the same mass of tissue in rads per second, rads per hour, or rads per day is

$$R_\gamma \,(\text{rad/sec}) = 5.92 \times 10^{-4} \, C \,(n_i E_i \phi_i)$$
$$R_\gamma \,(\text{rad/hr}) \;= 2.13 \, C \; (n_i E_i \phi_i)$$
$$R_\gamma \,(\text{rad/day}) = 51.2 \, C \; (n_i E_i \phi_i)$$

where n_i is the fractional emission or number of photons with energy E_i emitted per disintegration, and ϕ_i is the absorbed fraction of the mass of tissue for photons of energy E_i.

The absorbed dose constant Δ_i is

$$\Delta_i = 2.13 \, n_i E_i$$

in units of (g-rad/μCi-hr). Hence, the expression for the instantaneous dose rate in rads per hour may be written

$$R_\gamma \,(\text{rad/hr}) = C \, \Delta_i \phi_i$$

For nuclides which emit photons with j different energies, the instantaneous dose rate is

$$R_\gamma \,(\text{rad/hr}) = C \sum_{i=1}^{j} \Delta_i \phi_i$$

Absorbed dose constants for selected nuclides are included in Table 23-2.

If the nuclide is eliminated exponentially from the mass of tissue, then the instantaneous dose rate from penetrating radiation may be written

$$R_\gamma \,(\text{rad/hr}) = \left(\sum_{i=1}^{i} \Delta_i \phi_i \right) C_{\max} \, e^{-\ln 2(t/T_{\text{eff}})}$$

where $C_{\max}$ is the maximum concentration attained when $t = 0$. An expression for the cumulative internal dose from penetrating radiation D_γ may be obtained by integrating the expression for R_γ.

TABLE 23-3 Absorbed Fractions ϕ for Uniform Distribution of Radioactivity in Ellipsoids[a,b]

Mass, kg	Photon Energy, MeV										
	0.020	0.030	0.040	0.060	0.080	0.100	0.160	0.364	0.662	1.460	2.750
2	0.702	0.407	0.317	0.131	0.072	0.099	0.113	0.112	0.134	0.099	0.096
4	0.762	0.485	0.325	0.176	0.127	0.133	0.144	0.148	0.155	0.133	0.120
6	0.795	0.529	0.345	0.206	0.157	0.155	0.163	0.170	0.173	0.155	0.134
8	0.815	0.560	0.366	0.228	0.179	0.172	0.178	0.187	0.189	0.171	0.147
10	0.830	0.583	0.385	0.247	0.196	0.185	0.190	0.200	0.202	0.183	0.156
20	0.868	0.649	0.460	0.308	0.250	0.233	0.234	0.245	0.250	0.223	0.187
30	0.884	0.685	0.508	0.346	0.284	0.265	0.264	0.273	0.280	0.248	0.207
40	0.893	0.709	0.541	0.374	0.310	0.290	0.287	0.294	0.301	0.267	0.222
50	0.900	0.727	0.567	0.397	0.332	0.312	0.305	0.312	0.317	0.282	0.235
60	0.905	0.741	0.585	0.416	0.351	0.330	0.321	0.327	0.330	0.294	0.247
70	0.909	0.753	0.600	0.432	0.368	0.346	0.335	0.340	0.341	0.306	0.257
80	0.912	0.763	0.613	0.446	0.383	0.361	0.348	0.351	0.351	0.316	0.265
90	0.916	0.772	0.624	0.459	0.397	0.374	0.359	0.362	0.360	0.325	0.274
100	0.918	0.780	0.634	0.471	0.409	0.386	0.369	0.371	0.368	0.334	0.283
120	0.924	0.793	0.652	0.492	0.431	0.407	0.388	0.389	0.384	0.350	0.298
140	0.929	0.804	0.670	0.511	0.450	0.425	0.405	0.405	0.399	0.364	0.310
160	0.933	0.814	0.688	0.528	0.466	0.440	0.421	0.420	0.415	0.378	0.321
180	0.937	0.821	0.708	0.544	0.480	0.454	0.436	0.433	0.432	0.391	0.331
200	0.940	0.828	0.729	0.559	0.491	0.466	0.451	0.446	0.449	0.403	0.340

[a]Tables 23-3 through 23-5 from G. Brownell, W. Ellett and A. Reddy. 1968. Absorbed fractions for photon dosimetry. Report of MIRD Committee. J. Nuclear Med. Suppl. 1.
[b]The principal axes of the ellipsoids are in the ratio of 1:1.8:9.27.

TABLE 23-4 Absorbed Fractions ϕ for Uniform Distribution of Radioactivity in Small Spheres and Thick Ellipsoids[a]

Mass, kg	Photon energy, MeV										
	0.020	0.030	0.040	0.060	0.080	0.100	0.160	0.364	0.662	1.460	2.750
0.3	0.684	0.357	0.191	0.109	0.086	0.085	0.087	0.099	0.096	0.092	0.077
0.4	0.712	0.388	0.212	0.121	0.096	0.093	0.097	0.108	0.108	0.099	0.083
0.5	0.731	0.412	0.229	0.131	0.104	0.099	0.104	0.116	0.117	0.104	0.089
0.6	0.745	0.431	0.244	0.140	0.111	0.105	0.111	0.122	0.124	0.109	0.093
1.0	0.780	0.486	0.289	0.167	0.135	0.125	0.130	0.142	0.144	0.125	0.106
2.0	0.818	0.559	0.360	0.212	0.173	0.160	0.162	0.174	0.173	0.153	0.127
3.0	0.840	0.600	0.405	0.245	0.201	0.188	0.186	0.197	0.195	0.174	0.143
4.0	0.856	0.629	0.438	0.271	0.222	0.209	0.205	0.216	0.213	0.190	0.156
5.0	0.868	0.652	0.464	0.294	0.241	0.227	0.222	0.231	0.228	0.204	0.167
6.0	0.876	0.671	0.485	0.312	0.258	0.241	0.236	0.245	0.240	0.216	0.177

[a]The principal axes of the small spheres and thick ellipsoids are in the ratios of $1:1:1$ and $1:0.667:1.333$.

TABLE 23-5 *Absorbed Fractions ϕ for Uniform Distribution of Radioactivity in Flat Ellipsoids[a]*

Mass, kg	Photon energy, MeV								
	0.020	0.030	0.040	0.060	0.080	0.100	0.160	0.662	2.75
0.3	0.627	0.306	0.164	0.090	0.075	0.072	0.078	0.084	0.062
0.4	0.654	0.334	0.179	0.098	0.081	0.079	0.085	0.095	0.069
0.5	0.674	0.356	0.192	0.106	0.087	0.085	0.090	0.103	0.074
0.6	0.690	0.374	0.204	0.112	0.092	0.090	0.095	0.109	0.079
1.0	0.731	0.423	0.243	0.134	0.109	0.106	0.112	0.128	0.093
2.0	0.779	0.492	0.305	0.173	0.140	0.133	0.140	0.154	0.112
3.0	0.803	0.533	0.344	0.200	0.162	0.154	0.159	0.171	0.125
4.0	0.820	0.564	0.372	0.221	0.181	0.171	0.174	0.185	0.136
5.0	0.833	0.588	0.394	0.238	0.197	0.185	0.187	0.197	0.146
6.0	0.844	0.698	0.414	0.254	0.211	0.198	0.197	0.209	0.156

[a]The principal axes of the flat ellipsoids are in the ratio of $1:0.5:2.0$.

$$D_\gamma \text{ (rad)} = 1.44 \left(\sum_{i=1}^{j} \Delta_i \phi_i \right) C_{max} T_{eff} (1 - e^{-\ln 2(t/T_{eff})})$$

with t and T_{eff} expressed in hours. If t and T_{eff} are expressed in days, then

$$D_\gamma \text{ (rad)} = 34.6 \left(\sum_{i=1}^{j} \Delta_i \phi_i \right) C_{max} T_{eff} (1 - e^{-\ln 2(t/T_{eff})})$$

The total absorbed dose delivered by penetrating radiation during complete elimination of the nuclide from the mass of tissue is obtained by letting t approach ∞ in the expression above. The exponential term approaches 0 and

$$D_\gamma \text{ (rad)} = 34.6 \left(\sum_{i=1}^{j} \Delta_i \phi_i \right) C_{max} T_{eff}$$

If the effective half-life for uptake of a nuclide is very short compared to the effective half-life for elimination, then this expression may be used to estimate the total dose delivered by penetrating radiation from the moment the nuclide is distributed in a mass of tissue until the time it is eliminated completely. If the uptake of the nuclide must be considered, and if the uptake may be described exponentially, then the expression for the total cumulative dose D_γ from penetrating radiation may be written

$$D_\gamma \text{ (rad)} = 34.6 \left(\sum_{i=1}^{j} \Delta_i \phi_i \right) C_{max} T_{eff} \left(1 - \frac{T_{up}}{T_{eff}} \right)$$

TABLE 23-6 *Absorbed Fractions for Uniformly Distributed Sources in Small Unit-Density Spheres Surrounded by Scattering Medium*[a]

Mass, g	Photon energy, MeV									
	0.030	0.040	0.050	0.080	0.100	0.140	0.364	0.662	1.460	2.750
1	0.050	0.023	0.011	0.009	0.009	0.010	0.011	0.011	0.010	0.008
2	0.064	0.080	0.014	0.012	0.012	0.012	0.014	0.014	0.012	0.010
4	0.081	0.038	0.019	0.016	0.015	0.016	0.018	0.018	0.016	0.013
6	0.092	0.043	0.022	0.018	0.017	0.018	0.020	0.020	0.018	0.014
8	0.103	0.049	0.024	0.020	0.020	0.020	0.023	0.023	0.020	0.016
10	0.111	0.054	0.027	0.022	0.021	0.022	0.025	0.024	0.021	0.017
20	0.139	0.070	0.035	0.029	0.027	0.028	0.031	0.031	0.027	0.022
40	0.174	0.090	0.046	0.037	0.036	0.036	0.039	0.038	0.033	0.027
60	0.230	0.121	0.064	0.050	0.048	0.048	0.053	0.052	0.045	0.035
80	0.286	0.152	0.079	0.064	0.061	0.061	0.066	0.065	0.056	0.046
100	0.306	0.165	0.087	0.070	0.067	0.066	0.072	0.070	0.061	0.050

[a]Tables 23-6 through 23-8 from W. Ellett and R. Humes. 1971. Absorbed fractions for small volumes containing photon-emitting radioactivity. Report of MIRD Committee. J. Nuclear Med. Suppl. 5.

TABLE 23-7 *Absorbed Fractions for Uniformly Distributed Sources in Small Unit-Density Ellipsoids Surrounded by Scattering Medium (Axes 1 : 2 : 4)*

Mass g	Photon energy, MeV									
	0.030	0.040	0.060	0.080	0.100	0.140	0.364	0.662	1.460	2.750
1	0.045	0.021	0.010	0.008	0.008	0.009	0.010	0.010	0.009	0.007
2	0.058	0.027	0.013	0.011	0.011	0.011	0.013	0.013	0.011	0.009
4	0.073	0.035	0.017	0.014	0.014	0.014	0.016	0.016	0.014	0.012
6	0.082	0.040	0.020	0.016	0.016	0.016	0.018	0.018	0.016	0.014
8	0.0	0.045	0.022	0.018	0.018	0.018	0.021	0.020	0.018	0.015
10	0.100	0.049	0.024	0.020	0.019	0.020	0.022	0.022	0.019	0.016
20	0.125	0.063	0.032	0.026	0.025	0.025	0.028	0.028	0.024	0.020
40	0.155	0.081	0.042	0.034	0.032	0.032	0.035	0.035	0.030	0.025
60	0.192	0.101	0.052	0.043	0.040	0.041	0.044	0.044	0.037	0.031
80	0.229	0.121	0.063	0.051	0.049	0.049	0.053	0.052	0.045	0.037
100	0.244	0.131	0.069	0.056	0.053	0.053	0.057	0.056	0.049	0.040

TABLE 23-8 *Absorbed Fractions for Uniformly Distributed Sources in Small Ellipsoids Surrounded by Scattering Medium (Axes 1:3:8)*

Mass, g	Photon Energy, MeV									
	0.030	0.040	0.060	0.080	0.100	0.140	0.364	0.662	1.460	2.750
1	0.041	0.019	0.009	0.008	0.007	0.008	0.009	0.009	0.008	0.007
2	0.049	0.023	0.011	0.009	0.009	0.010	0.011	0.011	0.010	0.008
4	0.063	0.030	0.015	0.012	0.012	0.012	0.014	0.014	0.012	0.010
6	0.071	0.034	0.017	0.014	0.013	0.014	0.016	0.016	0.013	0.012
8	0.079	0.038	0.019	0.016	0.015	0.016	0.018	0.018	0.015	0.013
10	0.085	0.042	0.021	0.017	0.017	0.017	0.019	0.019	0.016	0.014
20	0.177	0.059	0.031	0.024	0.024	0.024	0.027	0.027	0.022	0.019
40	0.158	0.080	0.041	0.034	0.032	0.033	0.036	0.036	0.031	0.025
60	0.179	0.093	0.048	0.039	0.037	0.038	0.041	0.041	0.035	0.029
80	0.197	0.104	0.054	0.045	0.042	0.043	0.046	0.045	0.039	0.032
100	0.212	0.113	0.060	0.049	0.046	0.046	0.050	0.049	0.043	0.035

Combined Internal β-γ Dose

An expression for the instantaneous dose rate $R_{\beta+\gamma}$ delivered to tissue by a nuclide which emits radiation absorbed both locally and remotely is obtained by adding the appropriate equations stated earlier. The result is

$$R_{\beta+\gamma} \text{ (rad/hr)} = C \left(2.13\, \bar{E}_\beta + \sum_{i=1}^{j} \Delta_i \phi_i \right)$$

Similarly, the expression for the total cumulative $(\beta + \gamma)$ dose is

$$D_{\beta+\gamma} \text{ (rad)} = C_{max}\, T_{eff} \left(1 - \frac{T_{up}}{T_{eff}} \right) \left(73.8\, \bar{E}_\beta + 34.6 \sum_{i=1}^{j} \Delta_i \phi_i \right)$$

with T_{eff} and T_{up} in days. If the half-life T_{up} for uptake may be neglected, then

$$D_{\beta+\gamma} \text{ (rad)} = C_{max}\, T_{eff} \left(73.8\, \bar{E}_\beta + 34.6 \sum_{i=1}^{j} \Delta_i \phi_i \right)$$

Example 23-4

What is the total absorbed dose delivered to the kidney of the standard man by 100 μCi of ^{197}Hg-labeled chlormerodrin? The kidney mass is 300 g in the standard man.

Chlormerodrin labeled with ^{197}Hg is eliminated from the kidney with an effective half-life of 0.21 days (Example 23-1). Data for the decay of ^{197}Hg follows.

Radiation	Energy E_i, MeV	Fractional emission n_i	$n_i E_i$, MeV
M internal conversion electron from γ_1	0.0746	0.203	0.015
M internal conversion electron from γ_2	0.1888	0.0005	—
L internal conversion electron from γ_1	0.0640	0.609	0.039
L internal conversion electron from γ_2	0.1782	0.0014	—
K internal conversion electron from γ_2	0.1108	0.0072	0.001
K internal conversion electron from γ_3	0.1873	0.0004	—
L_α x-rays	0.0097	0.252	0.002
Auger electrons			
KLL (electron transition L to K with emission Auger electron from L shell)	0.0540	0.0191	0.001
KLX (electron transition L to K with emission Auger electron from shell above L)	0.0646	0.0108	0.001
KXY (electron transition from shell above L to K with emission Auger electron from shell above L)	0.0752	0.0018	
LMM (electron transition M to L with emission Auger electron from M shell)	0.0079	0.903	0.007
MXY (electron transition shell above M to M with Auger electron from shell above M)	0.0027	2.72	0.007

Radiation	Energy E_i, MeV[a]	Fractional emission $n_i{}^a$	Δ_i, $\dfrac{\text{g-rad}}{\mu\text{Ci-hr}}$	Absorbed fraction $\phi_i{}^b$
γ_1	0.0773	0.186	0.0306	0.088
γ_2	0.1915	0.0090	0.0037	0.0675
γ_3	0.2680	0.0010	0.0006	0.0691
K_{α_1} x-rays	0.0688	0.363	0.0532	0.095
K_{α_2} x-rays	0.0670	0.199	0.0284	0.092
K_{β_1} x-rays	0.0780	0.126	0.0209	0.088
K_{β_2} x-rays	0.0807	0.0338	0.0058	0.083
L_β x-rays	0.0115	0.236	0.0058	0.888
L_γ x-rays	0.0134	0.0317	0.0009	0.833

[a]From C. Dillman. 1969. Radionuclide decay values and nuclear parameters for use in radiation-dose evaluation. Report of the medical internal radiation dose committee. J. Nuclear Med. Suppl. 2, pamphlet 4.
[b]From W. Snyder, M. Ford, G. Warner, and H. Fisher. 1969. Estimates of absorbed fractions for monoenergetic photon sources uniformly distributed in various organs of a heterogeneous phantom. Report of the medical internal radiation dose committee. J. Nuclear Med. Suppl. 3, pamphlet 5.

Computation of $\bar{E}_\beta$ (see Table 23-2):
$$\bar{E}_\beta = \Sigma n_i E_i = 0.073 \text{ MeV}$$

$$D_{\beta+\gamma} = C_{\max} T_{\text{eff}} \left(73.8\,\bar{E}_\beta + 34.6 \sum_{i=1}^{j} \Delta_i \phi_i\right)$$

$$= \frac{100\,\mu\text{Ci}}{300\,\text{g}} (0.21 \text{ days})\{(73.8(0.073 \text{ MeV}) + 34.6[(0.0306)(0.088)$$

$$+ (0.0037)(0.0675) + (0.0006)(0.0691) + (0.0532)(0.095)$$

$$+ (0.0284)(0.092) + (0.0209)(0.088) + (0.0058)(0.083)$$

$$+ (0.0058)(0.888) + (0.0009)(0.833)]\}$$

$$= 0.422 \text{ rad}$$

$$= 422 \text{ mrad}$$

Example 23-5

What activity of ^{131}I should be administered orally for a dose of 7000 rad to the thyroid of a patient with Graves disease (hyperthyroidism). The mass of the thyroid is estimated as 40 g, the uptake of iodine is 75%, and the effective half-life is 2.3 days. The thyroid is assumed to be a small unit density ellipsoid (axes $1:2:4$). Data for the decay of ^{131}I are:

$$\bar{E}_\beta = 0{\cdot}188 \text{ MeV}$$

Radiation	Energy E_i, MeV[a]	Fractional emission n_i	Δ_i, g-rad/μCi-hr	Absorbed fraction ϕ_i[b]
γ_1	0.0456	0.0294	0.0029	0.050
γ_2	0.1640	0.0001	—	0.032
γ_3	0.1772	0.0014	0.0005	0.032
γ_4	0.2843	0.0475	0.0288	0.034
γ_5	0.3258	0.0017	0.0012	0.034
γ_6	0.3645	0.833	0.6465	0.035
γ_7	0.5030	0.0032	0.0034	0.035
γ_8	0.6370	0.0687	0.0932	0.035
γ_9	0.7329	0.0159	0.0245	0.035
K_{α_1} x-rays	0.0298	0.0252	0.0016	0.155
K_{α_2} x-rays	0.0295	0.0130	0.0008	0.155
K_{β_1} x-rays	0.0336	0.0070	0.0005	0.125
K_{β_2} x-rays	0.0346	0.0015	0.0001	0.118

[a]From L. Dillman. 1969. Radionuclide decay schemes and nuclear parameters for use in radiation-dose estimation. Report of the medical internal radiation dose committee. J. Nuclear Med. Suppl. 2, pamphlet 4.
[b]From Table 23-7.

$$D_{\beta+\gamma} = C_{\max} T_{\text{eff}} \left[73.8\, \bar{E}_\beta + 34.6 \sum_{i=1}^{j} \Delta_i \phi_i \right]$$

$$7000 \text{ rads} = \frac{x}{40 \text{ g}} (2.3 \text{ days})\{73.8(0.188 \text{ MeV}) + 34.6[(0.0029)(0.050)$$

$$+ (0.0005)(0.032) + (0.0288)(0.034) + (0.0012)(0.034)$$

$$+ (0.6465)(0.035) + (0.0034)(0.035) + (0.0932)(0.035)$$

$$+ (0.0245)(0.035) + (0.0016)(0.155) + (0.0008)(0.155)$$

$$+ (0.0005)(0.125) + (0.0001)(0.188)]\}$$

$$= 0.8544\, x$$

$$x = 8200\, \mu\text{Ci}$$

$$= 8.2 \text{ mCi}$$

The maximum activity in the thyroid must be 8.2 mCi. Since the percent uptake is 75, the amount of iodine to be administered orally is

$$\frac{8.2 \text{ mCi}}{0.75} = 10.9 \text{ mCi}$$

PROBLEMS

1. ^{24}Na administered as NaCl is distributed rapidly throughout the body and eliminated with a biologic half-life of 30 days. Compute the instantaneous β dose rate, the β dose delivered over the first 24 hr, and the cumulative β dose after intravenous administration of 25 μCi of ^{24}Na as NaCl to a 70 kg man. The average β energy is 0.55 MeV for this nuclide.

2. For the data in Problem 1, compute the corresponding γ dose rate, and 24 hr and cumulative doses for the two principal γ-rays from ^{24}Na.

Radiation	Energy	Fractional emission	Absorbed fraction in whole body
γ_1	1.3085	0.999	0.307
γ_2	2.7539	0.999	0.267

3. What are the total instantaneous dose rates, the total dose delivered after 24 hr, and the total cumulative dose delivered by 25 μCi of ^{24}Na as NaCl administered to a 70 kg man?

4. The disappearance of ^{35}S in the skin was determined by counting serial biopsies after injection of a tracer dose of Na_2 $^{35}SO_4$ in the skin. From the following data, determine the effective and biologic half-life of ^{35}S in the skin.

counts/min/g	6521	5543	4891	4108	3523	2606	1892	1043
time, days	0	5	10	15	20	30	40	60

REFERENCES

[1] International Commission on Radiological Protection. 1969. Report of committee II; permissible dose for internal radiation. ICRP Publ. 2. Pergamon Press, New York.

[2] Smith, E. 1968. Radiation dosimetry. *In* H. Wagner (ed.), Principles of nuclear medicine. W. B. Saunders Co., Philadelphia. p. 742.

[3] Reports of MIRD Committee. 1968–1971. J. Nuclear Med., Suppl. 1–5.

24

Guidelines for Radiation Protection

During the first few years after their discovery, x-rays and radioactive materials were used with little knowledge of their biologic effects and little concern for the possible harmful consequences of exposure to radiation. These consequences became apparent after a few years, however, and stimulated the establishment of radiation protection guidelines to permit the continued use of radiation sources without subjection of individuals or the general population to unacceptable risks.

Organizations for Radiation Protection

The responsibility for recommending limits for exposure of persons to radiation has been assumed by advisory groups composed of persons experienced in the use of radiation sources and knowledgeable about the biologic effects of radiation exposure. The establishment of recommended limits for radiation exposure was initiated in 1922 by the American Roentgen Ray Society and the American Radium Society. In 1928, the International Congress of Radiology formed the International Commission on Radiological Protection, referred to commonly as the ICRP. This organization is recognized today as the international authority for the safe use of sources of ionizing radiation. In 1929, the National Council on Radiation Protection (NCRP), at that time entitled the U.S. Advisory Committee on X-Ray and Radium Protection, was formed to interpret and implement recommendations of the ICRP for the United States. In

311

this country, regulation and control of radioactive and fissionable materials are the responsibilities of the U.S. Atomic Energy Commission. In some states, this responsibility has been delegated to the state departments of health.

Radiation Protection Guidelines

In recommending upper limits for exposure of persons to ionizing radiation, the ICRP assumes that a population is comprised of various groups of persons. For persons whose occupations or research interests require exposure to radiation, dose-limiting recommendations are[1]

Gonads, red bone marrow, lenses of the eyes, and whole body (in case of uniform irradiation)	5 rem/yr
Skin	15 rem/yr
Hands	75 rem/yr
Forearms	30 rem/yr
All other organs	15 rem/yr
Fertile women (with respect to fetus)	0.5 rem in the gestation period

Recommended dose limits include contributions from radiation sources both inside and outside the body, but exclude contributions from medical exposure and background radiation. The total accumulated dose equivalent in rems received by an occupationally exposed person should not exceed $5(N-18)$, where N is the age of the person in years. A working area where the yearly dose equivalent might reach 1.5 rem to the whole body of an individual should be considered a controlled area and should be supervised by a radiation safety officer. Persons working in a controlled area should carry one or more personal monitors for estimation of their whole-body dose. Personal monitors include film badges, thermoluminescent dosimeters, and pocket ionization chambers.

A variety of problems arise when recommendations regarding radiation exposure are applied to particular individuals. For occupational exposure, some of these problems involve:

(a) Persons with an unknown exposure history. For the period of unknown exposure, these persons are assumed to have received a maximum permissible dose for each year of age beyond 18.

(b) Persons whose exposures were restricted by dose limits established in earlier years. Since earlier limits were higher than those recommended now, these persons may have accumulated a whole-body or critical organ dose greater than that given by the expression

$5(N-18)$. The yearly dose received by these persons should be reduced below 5 rem/yr until the total accumulated dose is less than $5(N-18)$.

(c) Persons beginning work at an age less than 18 years. For these persons, the dose to the gonads, red bone marrow, lenses, and whole body should not exceed 5 rem/yr and the dose accumulated by age 30 should not exceed 60 rem.

(d) Women of reproductive capacity. These women should work in an environment where the abdominal dose is limited to 1.3 rem during any period of 13 weeks. Under this limitation, the dose to an embryo will be less than 1 rem during the first 2 months of organogenesis, the period during which the embryo is most sensitive to radiation.

(e) Pregnant women. When a pregnancy has been determined, conditions of employment should be arranged to ensure that the dose to the fetus is not more than 0.5 rem during the remaining period of pregnancy.

Dose limits for members of the public who are exposed occasionally to radiation are 1/10 of the maximum permissible dose equivalents for occupationally exposed persons. For example, the recommended dose limit is 0.5 rem/yr for the gonads, red bone marrow, lenses of the eyes, and the whole-body.

The average radiation dose to an entire population is a reflection not only of the radiation dose received by individual members of the population, but also of the total number of persons exposed in the population. Recommended limits for exposure of a population are based primarily on concern for the genetic effects of radiation exposure. The genetic dose to a population is the dose which, if received by each member of the population from conception to the mean age of childbearing (assumed to be 30 years of age), would introduce into the population a genetic burden equal to that furnished by individuals in the population who actually have been exposed to radiation. A permissible genetic dose is the dose which furnishes a genetic burden acceptable to the population. The genetic dose should be kept as low as possible and should not exceed 5 rem for all sources of ionizing radiation except background and medical exposure.

Body Burdens and Critical Organs

The body burden of a particular radioactive nuclide in a particular individual is the activity of the nuclide present in the individual's body. The maximum permissible body burden is the activity of a particular nuclide which delivers a maximum permissible dose to the whole body or to one or more organs in the body. Maximum permissible body burdens are computed with the assumption that the particular nuclide is the only radioactive nuclide in the body. If more than one nuclide is present, then the permissible body burden for each nuclide must be reduced accordingly. A nuclide retained in the body at a level lower than the maximum permissible body burden should cause observable biologic damage to the individual or his progeny in only the rarest of instances.

For a radioactive nuclide of a bone-seeking element (such as calcium or strontium), the maximum permissible body burden is the number of microcuries required to deliver to bone a dose in rems equal to that provided by 0.1 μCi of ^{226}Ra and its decay products. For a nuclide which is not a bone seeker, the determination of the maximum permissible body burden requires identification of one or more critical organs for the nuclide. The selection of a critical organ or organs requires evaluation of many factors, including: (a) the concentration of the nuclide in different organs; (b) the sensitivity of different organs to ionizing radiation; (c) the importance of different organs to the health of the individual; and (d) the dose to the whole body and to the organs irradiated during intake and elimination of the nuclide. In most cases, the concentration of the nuclide is the most influential factor in selecting a critical organ or organs.

If a radioactive nuclide is distributed fairly uniformly throughout the body, then the whole body may be selected as the critical organ. In this situation, the maximum permissible body burden is the activity present continuously in the body which delivers a dose equivalent of 5 rem/yr to the whole body. The nuclides ^{35}S and ^{127m}Te concentrate in the testes and are assigned a maximum permissible body burden which delivers a dose of 5 rem/yr to this organ. For nuclides which concentrate in the thyroid or in abdominal organs, the maximum permissible body burdens are the activities which deliver 15 rem/yr to these organs.

Maximum Permissible Concentrations

Restricting the uptake of radioactive materials into the body is the most effective method for reducing the hazards associated with radioactivity deposited internally. The uptake may be restricted by controlling the concentration of radioactive nuclides in the air and water available for human consumption. Maximum permissible concentrations in air and water have been established for exposures of 40 hr/week and 168 hr/week. A radioactive nuclide in air or water at a concentration less than the maximum permissible concentration should result in a body burden for the nuclide which is less than the maximum permissible body burden. Selected maximum permissible body burdens and maximum permissible concentrations in air and water are listed in Table 24-1.

TABLE 24-1 Maximum Permissible Body Burdens and Maximum Permissible Concentrations for Selected Radioactive Nuclides in Soluble and Insoluble Form

Radioactive nuclide and mode of decay	Organ of reference	Maximum permissible body burden, μCi	Maximum permissible concentration, μCi/cm^3			
			40 hr week		168 hr week	
			$(MPC)_w$	$(MPC)_a$	$(MPC)_w$	$(MPC)_a$
^{3}H(^{3}H$_2$O)(β^-) (soluble)	Body tissue	10^3	0.1	2×10^{-5}	0.03	5×10^{-6}
	Total body	2×10^3	0.2	2×10^{-5}	0.05	7×10^{-6}
^{3}H (immersion)	Skin			2×10^{-3}		4×10^{-4}
^{14}C(CO$_2$)(β^-)(soluble)	Fat	200	0.02	4×10^{-6}	8×10^{-3}	10^{-6}
	Total body	400	0.03	5×10^{-6}	0.01	2×10^{-6}
	Bone	400	0.04	6×10^{-6}	0.01	2×10^{-6}
^{14}C (immersion)	Total body			5×10^{-5}		10^{-5}
^{32}P (β^-) (soluble)	Bone	6	5×10^{-4}	7×10^{-8}	2×10^{-4}	2×10^{-8}
	Total body	30	3×10^{-3}	4×10^{-7}	9×10^{-4}	10^{-9}
	GI(LLI)	—	3×10^{-3}	6×10^{-7}	9×10^{-4}	2×10^{-7}
	Liver	50	5×10^{-3}	6×10^{-7}	2×10^{-3}	2×10^{-7}
	Brain	300	0.02	3×10^{-6}	8×10^{-3}	10^{-6}
(insoluble)	Lung			8×10^{-8}		3×10^{-8}
	GI(LLI)		7×10^{-4}	10^{-7}	2×10^{-4}	4×10^{-8}

REFERENCE

[1] National Council on Radiation Protection and Measurements. 1971. Recommendations of the National Council on Radiation Protection and Measurements, basic radiation protection criteria. NCRP Report 39. Washington, D.C.

25
Radiation Safety

In every laboratory in which radioactive nuclides are used, equipment should be available for monitoring the exposure of persons to radiation and for checking for radioactive contamination of personnel and facilities. Also, procedures should be established and followed to minimize the possibility of external contamination and the likelihood of ingestion, inhalation, and absorption of radioactive material. A few procedures which are generally applicable are described here; other procedures related to the specific requirements of a particular laboratory may also be required.

Personnel Monitors

Persons working with or near sources of penetrating radiation should carry personnel dosimeters which indicate the exposure of the persons to radiation. The dosimeter used most often is the film badge, comprised of two or more small x-ray films enclosed within a light-tight envelope and plastic case. The badge is worn from 1 to 4 weeks on the trunk of the body, usually at the level of the belt or shirt pocket. Badges in special holders may be worn also on the wrist as a wrist badge or on a finger as a ring badge. After the badge has been worn for the prescribed period, each film is processed and its optical density is compared to that for similar films receiving known exposures. From these comparisons, the radiation dose may be estimated for the film badge and, presumably, for the individual wearing the badge. Small metal filters in the plastic case produce

317

variations in the optical density of the films which permit some distinction among different types and energies of radiation which may have contributed to the exposure. Types of radiation which can be monitored with a film badge include x- and γ-rays, neutrons, high-energy electrons, and β particles.

In some institutions, thermoluminescent dosimeters, usually LiF or CaF, have replaced x-ray film as a personnel monitor. Exposure of a thermoluminescent dosimeter to radiation results in the trapping of electrons in energy levels above those occupied normally. When the dosimeter is heated, these electrons are liberated from the traps. As the electrons return to their normal energy levels, visible light is released. The amount of light released is measured with a photomultiplier tube and is proportional to the exposure of the thermoluminescent dosimeter to radiation[1].

Occasionally, other types of personnel monitors (such as pocket ionization chambers) are used in association with film badges or thermoluminescent dosimeters. Some monitors produce an audible warning when carried into an area where the exposure rate is unusually high.

Area Monitors

An area monitor may be placed in a strategic location in a radioisotope laboratory to monitor the radiation levels continuously at that location. Usually, the location for an area monitor is near the site where radioactive nuclides are handled or stored. Most monitors provide an audible alarm when the radiation level exceeds a preset value. In most area monitors, the radiation-sensitive component is an ionization chamber. A typical area monitor is shown in Figure 25-1.

Survey Instruments

Two types of portable survey instruments are in widespread use. One type employs a GM detector to detect the presence of radioactive contamination in working areas or on persons handling radioactive materials. A GM survey meter equipped with a thin end-window or a thin wall ($1-2$ mg/cm^2) detector is useful for detecting contamination by most radioactive nuclides (with the notable exception of tritium). Unless this instrument is calibrated carefully and extensively for one or more nuclides, it should not be used for measurements of exposure rates in the vicinity of γ-emitting nuclides. A typical GM survey meter with a thin window detector is illustrated in Figure 25-2.

The second common type of portable survey instrument is equipped

Figure 25-1 A typical area monitor. (Courtesy of Victoreen Instrument Co.)

with an ionization chamber, and is used to measure exposure rates in the vicinity of sources of x- or γ-radiation. The "cutie pie" illustrated in Figure 25-3 is a typical portable survey instrument. Exposure rates from 1 mR/hr to 100 R/hr are measurable with the instrument illustrated in Figure 25-3.

Posting Requirements

Any room or area in which significant quantities of radioactive isotopes are used or stored should be posted with a sign indicating the presence of radioactive material. If the dose rate to a person in the area might be as high as 5 mrem/hr or 100 mrem in 5 consecutive days, then the area should be posted with a sign which designates the area as a "Radiation Area." If the dose rate might exceed 100 mrem/hr, then the area should be labeled a "High Radiation Area." A sign with the words "Airborne Radio-activity Area" should be posted at the entrance to any area where the airborne concentration of radioactivity approaches the maximum permissible concentration. An area containing an x-ray machine should be posted with a sign indicating "Caution: X-Ray." Various signs which satisfy posting requirements are illustrated in Figure 25-4.

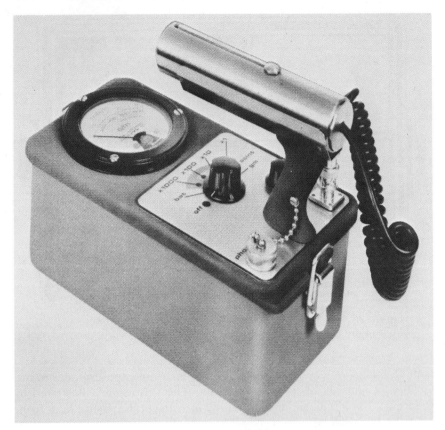

Figure 25-2   A thin window GM survey meter.

Containers for radioactive materials should be labeled with the radiation caution symbol together with data concerning the nuclide, the activity, the date of assay, and the name of the user. Containers used transiently in laboratory procedures need not be labeled, provided that the user is present and the containers are cleaned after use.

Housekeeping

Procedures for using radioactive materials without causing contamination of laboratory facilities resemble closely the procedures for good laboratory technique in general. The working area should be kept neat, and glassware and equipment should be cleaned immediately after use. Equipment and supplies not required for an experiment should be removed from the working area before the experiment is begun. Manipulations involving

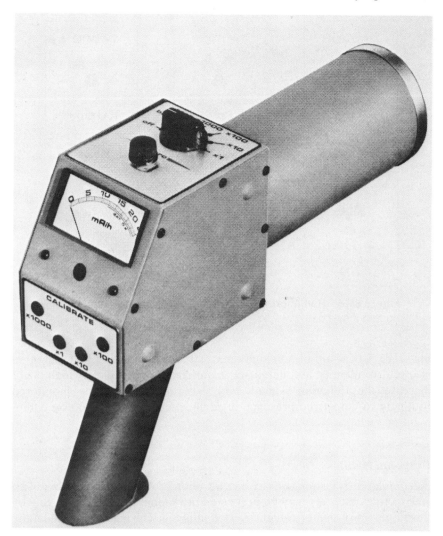

Figure 25-3 Cutie pie portable survey meter.

radioactive materials should be confined to as small an area as possible, and should be performed in a tray lined with absorbent paper with an impermeable backing. The laboratory bench where the manipulations are performed also should be covered with absorbent paper. All radioactive samples not in use should be stored in an appropriate container. Procedures involving the possible release of radioactive material into the air should be conducted in a ventilated fumehood with surfaces of nonporous

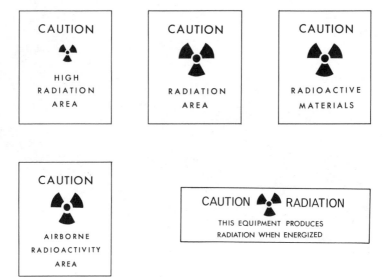

Figure 25-4   Radiation warning signs for areas where a radiation hazard may exist.

material. A trial run of an experiment without radioactive material often is useful to identify and eliminate unanticipated difficulties which might increase the radiation hazard. The laboratory should be monitored routinely to detect the presence of radioactive contamination on equipment or facilities.

Personal Habits

Accumulation of radioactivity in the bodies of persons working with radioactive materials may be prevented or minimized most effectively by development of personal habits which minimize the ingestion, inhalation, and absorption of radioactivity. For example, eating, drinking, smoking, and placing items in the mouth should not be permitted in a radioisotope laboratory. Radioactive samples should be pipetted with a device such as a propipet and not by mouth suction. Clothing appropriate for the types and quantities of radioactive nuclides used, and for the degree of contamination hazard present, should be worn at all times in the laboratory. Protective clothing should never be less than a laboratory coat and, when contamination of the hands is possible, plastic gloves. Before leaving the laboratory, each person should monitor his hands and other areas of his body which may possibly be contaminated with radioactivity.

When working near sources of penetrating radiation, persons should remember that their exposure: (1) increases with time spent near the sources; (2) decreases with increased distance from the sources; and (3) decreases with increased shielding around the sources.

Area Decontamination

If radioactive material has been spilled in an area, or if radioactive contamination of laboratory facilities has been detected, measures should be taken immediately to protect personnel in the laboratory and to confine the radioactivity. If the spill or contamination is extensive, then the steps below might be followed. These procedures may be relaxed somewhat if the contamination or spill is less extensive.

To confine the radioactivity:

(a) Close doors and windows.
(b) Turn off fans, air conditioners, forced air heaters, etc.
(c) Close ventilation ducts.
(d) Vacate room, leaving outer garments at door.
(e) Lock door and seal with masking tape.

To remove the radioactivity:

(a) Provide protective clothing and, if necessary, protective breathing apparatus to all persons involved in decontamination.
(b) Monitor area to determine the type and extent of contamination.
(c) Decontaminate by scrubbing contaminated facilities with a detergent solution, always washing toward the center of the contaminated area to retain the activity. Monitor frequently and thoroughly during decontamination. For nuclides such as ^3H which cannot be monitored with a portable survey meter, obtain wipe smears with dampened filter paper and measure the activity on the smears with a liquid scintillation counter.
(d) Monitor all personnel before their dismissal to uncontaminated areas.

Personnel Decontamination

When contamination of the skin is detected, a procedure similar to that outlined here should be followed:

(a) Wash contaminated area for 2–3 min with soap and water, starting with hands if they are contaminated. Use warm water and lather well, working lather under fingernails and other regions difficult to clean. Do not allow lather to drain upon uncontaminated areas. Repeat up to four times, or until monitoring indicates that contamination has been removed.

(b) If (a) is insufficient, repeat the procedure twice, using a brush with firm bristles and monitoring. Do not irritate or break the skin.

(c) If (a) or (b) is adequate, apply hand cream.

(d) If (a) or (b) is insufficient, apply a mixture of titanium dioxide and hand cream over the area for 2 min, using enough water to keep the paste moist. Rinse and wash thoroughly with soap and water. Monitor and repeat if necessary.

(e) If (d) is adequate, apply hand cream.

(f) If (d) is insufficient, obtain professional assistance.

Waste Disposal

Most of the radioactive waste from biomedical laboratories is "low-level waste" (that is, waste with relatively little activity), and can be released by disposal into the air, water or ground. A particular radioactive nuclide may be released into the sanitary sewage system, for example, provided that the $(MPC)_w$ for the nuclide is not exceeded after dilution of the radioactive waste by the sewage effluent from the building. Nuclides with short half-lives may be stored for decay before disposal, if an adequate facility for storage is available.

Solid wastes containing a particular radioactive nuclide may be incinerated if the concentration of the nuclide at the point of release into the air does not exceed the $(MPC)_a$ for the nuclide. The ash remaining after incineration must be monitored and retained as radioactive waste if the activity exceeds an acceptable level. In some locations, small quantities of radioactive waste may be buried in sanitary land–fill disposal sites.

Long-lived waste which is not disposable by incineration, dilution in the sewage effluent, or local burial, must be shipped to a burial site approved for disposal of radioactive waste. These sites are located in remote areas throughout the country.

Record Keeping

The maintenance of complete and current records is required whenever radioactive materials are used. These records should include the following data:

(a) A description of all radioactive samples acquired by the laboratory.

(b) A description of all radioactive waste released by the laboratory and the method of release.

(c) A record of the total activity in possession for each nuclide used in the laboratory.

(d) Results of all monitoring procedures for facilities and personnel in the laboratory.

(e) Results of all tests for leakage of radioactive material from sealed sources. Leak tests on all sealed sources should be performed at least once every 6 months by a procedure capable of detecting the transfer of 0.005 μCi to the leak test material. If more than 0.005 μCi is removed from a source during a leak test, the source should be withdrawn from use and disposed or returned to the manufacturer for repair.

(f) A cumulative record of the radiation exposure of each person in the laboratory, as reflected in reports of film badge readings or data from other types of personnel monitors.

(g) A report of the calibration of all survey instruments used in the laboratory.

REFERENCE

[1] Auxier, J., K. Becker, and E. Robinson (ed.). 1968. Proceedings Second International Conference on Luminescence Dosimetry, CONF-680920 Clearinghouse for Fedral Scientific and Technical Information, Springfield, Va.

Appendixes

Appendix 1

Derivation of the expressions $N = N_0 e^{-\lambda t}$ **and** $A = A_0 e^{-\lambda t}$

The expression

$$A = \lambda N \text{ (Chapter 2)}$$

may be restated as

$$\frac{dN}{dt} = -\lambda N$$

where dN/dt represents the rate of decay A of the sample, and the negative sign indicates that the number of radioactive atoms in the sample is decreasing. From this expression

$$\frac{dN}{N} = -\lambda \, dt$$

Over a finite interval of time from $t = 0$ to t, the number of atoms changes from N_0 to N.

$$\int_{N_0}^{N} \frac{dN}{N} = -\lambda \int_{0}^{t} dt$$

$$\ln N \, \Big]_{N_0}^{N} = -\lambda t \, \Big]_{0}^{t}$$

$$\ln N - \ln N_0 = -\lambda t$$

$$\ln \frac{N}{N_0} = -\lambda t$$

$$e^{\ln(N/N_0)} = e^{-\lambda t}$$

But, $e^{\ln(N/N_0)} = N/N_0$

$$\frac{N}{N_0} = e^{-\lambda t}$$

$$N = N_0 \, e^{-\lambda t}$$

If this expression is multiplied by λ

$$\lambda N = \lambda N_0 \, e^{-\lambda t}$$

But $A = \lambda N$ and $A_0 = \lambda N_0$

$$A = A_0\, e^{-\lambda t}$$

Derivation of the expression $\lambda = \ln 2/T_{1/2}$

The number of radioactive atoms (or the activity) of a sample decreases from N_0 to $1/2 N_0$ (or A_0 to $1/2 A_0$) during an interval of time equal to the half life $T_{1/2}$. Hence, from

$$A = A_0\, e^{-\lambda t}$$

$$1/2 A_0 = A_0\, e^{-\lambda T_{1/2}}$$

$$1/2 = e^{-\lambda T_{1/2}}$$

$$\ln 1/2 = \ln (e^{-\lambda T_{1/2}})$$

But $\ln (e^{-\lambda T_{1/2}}) = -\lambda T_{1/2}$

$$\ln 1/2 = -\lambda T_{1/2}$$

$$\ln 1 - \ln 2 = -\lambda T_{1/2}$$

Since $\ln 1 = 0$,

$$-\ln 2 = -\lambda T_{1/2}$$

$$\ln 2 = \lambda T_{1/2}$$

$$\lambda = \frac{\ln 2}{T_{1/2}}$$

Derivation of the expression $I = I_0 e^{-\mu x}$

The rate dI/dx at which photons are removed from an x- or γ-ray beam may be described as

$$\frac{dI}{dx} = -\mu I$$

where I represents the number of photons in the beam, and μ is the attenuation coefficient of the medium for the photons (Chapter 4). From this expression

$$\frac{dI}{I} = -\mu\, dx$$

Over a finite thickness x of the medium, the number of photons changes from I_0 to I.

$$\int_{I_0}^{I} \frac{dI}{I} = -\mu \int_{0}^{x} dx$$

$$\ln I \Big]_{I_0}^{I} = -\mu x \Big]_{0}^{x}$$

$$\ln I - \ln I_0 = -\mu x$$

$$\ln \frac{I}{I_0} = -\mu x$$

$$e^{\ln(I/I_0)} = e^{-\mu x}$$

But $e^{\ln(I/I_0)} = I/I_0$

$$\frac{I}{I_0} = e^{-\mu x}$$

$$I = I_0\, e^{-\mu x}$$

Derivation of the expression $HVL = \ln 2/\mu$

The number of x- or γ-ray photons decreases from I_0 to $1/2I_0$ as a narrow monoenergetic beam traverses a medium of thickness equal to the half-value layer HVL. Hence, from

$$I = I_0\, e^{-\mu x}$$

$$1/2I_0 = I_0\, e^{-\mu(\text{HVL})}$$

$$1/2 = e^{-\mu(\text{HVL})}$$

$$\ln 1/2 = \ln e^{-\mu(\text{HVL})}$$

But $\ln e^{-\mu(\text{HVL})} = -\mu(\text{HVL})$

$$\ln 1/2 = -\mu(\text{HVL})$$

$$\ln 1 - \ln 2 = -\mu(\text{HVL})$$

Since $\ln 1 = 0$

$$-\ln 2 = -\mu(\text{HVL})$$

$$\ln 2 = \mu(\text{HVL})$$

$$\text{HVL} = \frac{\ln 2}{\mu}$$

Appendix 2

Exponential Quantity e Raised to Selected Negative Powers, e^{-x}

X	0.00	0.01	0.02	0.03	0.04	0.05	0.06	0.07	0.08	0.09
0.0	1.000	0.9900	0.9802	0.9704	0.9608	0.9512	0.9418	0.9324	0.9231	0.9139
0.1	0.9048	0.8958	0.8869	0.8781	0.8694	0.8607	0.8521	0.8437	0.8353	0.8270
0.2	0.8187	0.8106	0.8025	0.7945	0.7866	0.7788	0.7711	0.7634	0.7558	0.7483
0.3	0.7408	0.7334	0.7261	0.7189	0.7118	0.7047	0.6977	0.6907	0.6839	0.6771
0.4	0.6703	0.6637	0.6570	0.6505	0.6440	0.6376	0.6313	0.6250	0.6188	0.6126
0.5	0.6065	0.6005	0.5945	0.5886	0.5827	0.5769	0.5712	0.5655	0.5599	0.5543
0.6	0.5488	0.5434	0.5379	0.5326	0.5273	0.5220	0.5169	0.5117	0.5066	0.5016
0.7	0.4966	0.4916	0.4868	0.4819	0.4771	0.4724	0.4677	0.4630	0.4584	0.4538
0.8	0.4493	0.4449	0.4404	0.4360	0.4317	0.4274	0.4232	0.4190	0.4148	0.4107
0.9	0.4066	0.4025	0.3985	0.3946	0.3906	0.3867	0.3829	0.3791	0.3753	0.3716
1.0	0.3679	0.3642	0.3606	0.3570	0.3535	0.3499	0.3465	0.3430	0.3396	0.3362
1.1	0.3329	0.3296	0.3263	0.3230	0.3198	0.3166	0.3135	0.3104	0.3073	0.3042
1.2	0.3012	0.2982	0.2952	0.2923	0.2894	0.2865	0.2837	0.2808	0.2780	0.2753
1.3	0.2725	0.2698	0.2671	0.2645	0.2618	0.2592	0.2567	0.2541	0.2516	0.2491
1.4	0.2466	0.2441	0.2417	0.2393	0.2369	0.2346	0.2322	0.2299	0.2276	0.2254
1.5	0.2231	0.2209	0.2187	0.2165	0.2144	0.2122	0.2101	0.2080	0.2060	0.2039
1.6	0.2019	0.1999	0.1979	0.1959	0.1940	0.1920	0.1901	0.1882	0.1864	0.1845
1.7	0.1827	0.1809	0.1791	0.1773	0.1755	0.1738	0.1720	0.1703	0.1686	0.1670
1.8	0.1653	0.1637	0.1620	0.1604	0.1588	0.1572	0.1557	0.1541	0.1526	0.1511
1.9	0.1496	0.1481	0.1466	0.1451	0.1437	0.1423	0.1409	0.1395	0.1381	0.1367
2.0	0.1353	0.1340	0.1327	0.1313	0.1300	0.1287	0.1275	0.1262	0.1249	0.1237
2.1	0.1225	0.1212	0.1200	0.1188	0.1177	0.1165	0.1153	0.1142	0.1130	0.1119
2.2	0.1108	0.1097	0.1086	0.1075	0.1065	0.1054	0.1043	0.1033	0.1023	0.1013
2.3	0.1003	*9926	*9827	*9730	*9633	*9537	*9442	*9348	*9255	*9163
2.4 0.0	9072	8982	8892	8804	8716	8629	8544	8458	8374	8291
2.5 0.0	8208	8127	8046	7966	7887	7808	7730	7654	7577	7502

x		0.0	0.1	0.2	0.3	0.4	0.5	0.6	0.7	0.8	0.9
2.6	0.0	7427	7353	7280	7208	7136	7065	6995	6925	6856	6788
2.7	0.0	6721	6654	6587	6522	6457	6393	6329	6266	6204	6142
2.8	0.0	6081	6020	5961	5901	5843	5784	5727	5670	5613	5558
2.9	0.0	5502	5448	5393	5340	5287	5234	5182	5130	5079	5029
3.0	0.0	4979	4929	4880	4832	4783	4736	4689	4642	4596	4550
3.1	0.0	4505	4460	4416	4372	4328	4285	4243	4200	4159	4117
3.2	0.0	4076	4036	3996	3956	3916	3877	3839	3801	3763	3725
3.3	0.0	3688	3652	3615	3579	3544	3508	3474	3439	3405	3371
3.4	0.0	3337	3304	3271	3239	3206	3175	3143	3112	3081	3050

X		0.0	0.1	0.2	0.3	0.4	0.5	0.6	0.7	0.8	0.9
3	0.0	4979	4505	4076	3688	3337	3020	2732	2472	2237	2024
4	0.0	1832	1657	1500	1357	1228	1111	1005	*9095	*8230	*7447
5	0.00	6738	6097	5517	4992	4517	4087	3698	3346	3028	2739
6	0.00	2479	2243	2029	1836	1662	1503	1360	1231	1114	1008
7	0.000	9119	8251	7466	6755	6112	5531	5004	4528	4097	3707
8	0.000	3355	3035	2747	2485	2249	2035	1841	1666	1507	1364
9	0.000	1234	1117	1010	*9142	*8272	*7485	*6773	*6128	*5545	*5017
10	0.0000	4540	4108	3717	3363	3043	2754	2492	2254	2040	1846

Exponential Quantity e Raised to Selected Positive Powers, e^x

x	0.00	0.01	0.02	0.03	0.04	0.05	0.06	0.07	0.08	0.09
0.0	1.000	1.010	1.020	1.031	1.041	1.051	1.062	1.073	1.083	1.094
0.1	1.105	1.116	1.127	1.139	1.150	1.162	1.174	1.185	1.197	1.209
0.2	1.221	1.234	1.246	1.259	1.271	1.284	1.297	1.310	1.323	1.336
0.3	1.350	1.363	1.377	1.391	1.405	1.419	1.433	1.448	1.462	1.477
0.4	1.492	1.507	1.522	1.537	1.553	1.568	1.584	1.600	1.616	1.632
0.5	1.649	1.665	1.682	1.699	1.716	1.733	1.751	1.768	1.786	1.804
0.6	1.882	1.840	1.859	1.878	1.896	1.916	1.935	1.954	1.974	1.994
0.7	2.014	2.034	2.054	2.075	2.096	2.117	2.138	2.160	2.181	2.203
0.8	2.226	2.248	2.270	2.293	2.316	2.340	2.363	2.387	2.411	2.435
0.9	2.460	2.484	2.509	2.535	2.560	2.586	2.612	2.638	2.664	2.691
1.0	2.718	2.746	2.773	2.801	2.829	2.858	2.886	2.915	2.945	2.974
1.1	3.004	3.034	3.065	3.096	3.127	3.158	3.190	3.222	3.254	3.287
1.2	3.320	3.353	3.387	3.421	3.456	3.490	3.525	3.561	3.597	3.633
1.3	3.669	3.706	3.743	3.781	3.819	3.857	3.896	3.935	3.975	4.015
1.4	4.055	4.096	4.137	4.179	4.221	4.263	4.306	4.349	4.393	4.437
1.5	4.482	4.527	4.572	4.618	4.665	4.712	4.759	4.807	4.855	4.904
1.6	4.953	5.003	5.053	5.104	5.155	5.207	5.259	5.312	5.366	5.419
1.7	5.474	5.529	5.585	5.641	5.697	5.755	5.812	5.871	5.930	5.989
1.8	6.050	6.110	6.172	6.234	6.297	6.360	6.424	6.488	6.554	6.619
1.9	6.686	6.753	6.821	6.890	6.959	7.029	7.099	7.171	7.243	7.316
2.0	7.389	7.463	7.538	7.614	7.691	7.768	7.846	7.925	8.004	8.085
2.1	8.166	8.248	8.331	8.415	8.499	8.585	8.671	8.758	8.846	8.935
2.2	9.025	9.116	9.207	9.300	9.393	9.488	9.583	9.679	9.777	9.875
2.3	9.974	10.07	10.18	10.28	10.38	10.49	10.59	10.70	10.80	10.91
2.4	11.02	11.13	11.25	11.36	11.47	11.59	11.70	11.82	11.94	12.06

X	0.0	0.1	0.2	0.3	0.4	0.5	0.6	0.7	0.8	0.9
2.5	12.18	12.30	12.43	12.55	12.68	12.81	12.94	13.07	13.20	13.33
2.6	13.46	13.60	13.74	13.87	14.01	14.15	14.30	14.44	14.59	14.73
3.0	20.09	20.29	20.49	20.70	20.91	21.12	21.33	21.54	21.76	21.98
3.1	22.20	22.42	22.65	22.87	23.10	23.34	23.57	23.81	24.05	24.29
3.2	24.53	24.78	25.03	25.28	25.53	25.79	26.05	26.31	26.58	26.84
3.3	27.11	27.39	27.66	27.94	28.22	28.50	28.79	29.08	29.37	29.67
3.4	29.96	30.27	30.57	30.88	31.19	31.50	31.82	32.14	32.46	32.79

X	0.0	0.1	0.2	0.3	0.4	0.5	0.6	0.7	0.8	0.9
3	20.09	22.20	24.53	27.11	29.96	33.12	36.60	40.45	44.70	49.40
4	54.60	60.34	66.69	73.70	81.45	90.02	99.48	109.9	121.5	134.3
5	148.4	164.0	181.3	200.3	221.4	244.7	270.4	298.9	330.3	365.0
6	403.4	445.9	492.7	544.6	601.8	665.1	735.1	812.4	897.8	992.3
7	1097	1212	1339	1480	1636	1808	1998	2208	2441	2697
8	2981	3295	3641	4024	4447	4915	5432	6003	6634	7332
9	8103	8955	9897	10938	12088	13360	14765	16318	18034	19930

Appendix 3

Natural Trigonometric Functions

Angle					Angle				
Degree	Radian	Sine	Cosine	Tangent	Degree	Radian	Sine	Cosine	Tangent
0°	0.000	0.000	1.000	0.000	46°	0.803	0.719	0.695	1.036
1°	0.017	0.018	1.000	0.018	47°	0.820	0.731	0.682	1.072
2°	0.035	0.035	0.999	0.035	48°	0.838	0.743	0.669	1.111
3°	0.052	0.052	0.999	0.052	49°	0.855	0.755	0.656	1.150
4°	0.070	0.070	0.998	0.070	50°	0.873	0.766	0.643	1.192
5°	0.087	0.087	0.996	0.088	51°	0.890	0.777	0.629	1.235
6°	0.105	0.105	0.995	0.105	52°	0.908	0.788	0.616	1.280
7°	0.122	0.122	0.993	0.123	53°	0.925	0.799	0.602	1.327
8°	0.140	0.139	0.990	0.141	54°	0.942	0.809	0.588	1.376
9°	0.157	0.156	0.988	0.158	55°	0.960	0.819	0.574	1.428
10°	0.175	0.174	0.985	0.176	56°	0.977	0.829	0.559	1.483
11°	0.192	0.191	0.982	0.194	57°	0.995	0.839	0.545	1.540
12°	0.209	0.208	0.978	0.213	58°	1.012	0.848	0.530	1.600
13°	0.227	0.225	0.974	0.231	59°	1.030	0.857	0.515	1.664
14°	0.244	0.242	0.970	0.249	60°	1.047	0.866	0.500	1.732
15°	0.262	0.259	0.966	0.268	61°	1.065	0.875	0.485	1.804
16°	0.279	0.276	0.961	0.287	62°	1.082	0.883	0.470	1.881
17°	0.297	0.292	0.956	0.306	63°	1.100	0.891	0.454	1.963
18°	0.314	0.309	0.951	0.325	64°	1.117	0.899	0.438	2.050
19°	0.332	0.326	0.946	0.344	65°	1.134	0.906	0.423	2.145
20°	0.349	0.342	0.940	0.364	66°	1.152	0.914	0.407	2.246
21°	0.367	0.358	0.934	0.384	67°	1.169	0.921	0.391	2.356
22°	0.384	0.375	0.927	0.404	68°	1.187	0.927	0.375	2.475
23°	0.401	0.391	0.921	0.425					

Degrees	Radians	sin	cos	tan
24°	0.419	0.407	0.914	0.445
25°	0.436	0.423	0.906	0.466
26°	0.454	0.438	0.899	0.488
27°	0.471	0.454	0.891	0.510
28°	0.489	0.470	0.883	0.532
29°	0.506	0.485	0.875	0.554
30°	0.524	0.500	0.866	0.577
31°	0.541	0.515	0.857	0.601
32°	0.559	0.530	0.848	0.625
33°	0.576	0.545	0.839	0.649
34°	0.593	0.559	0.829	0.675
35°	0.611	0.574	0.819	0.700
36°	0.628	0.588	0.809	0.727
37°	0.646	0.602	0.799	0.754
38°	0.663	0.616	0.788	0.781
39°	0.681	0.629	0.777	0.810
40°	0.698	0.643	0.766	0.839
41°	0.716	0.658	0.755	0.869
42°	0.733	0.669	0.743	0.900
43°	0.751	0.682	0.731	0.933
44°	0.768	0.695	0.719	0.966
45°	0.785	0.707	0.707	1.000

Degrees	Radians	sin	cos	tan
69°	1.204	0.934	0.358	2.605
70°	1.222	0.940	0.342	2.747
71°	1.239	0.946	0.326	2.904
72°	1.257	0.951	0.309	3.078
73°	1.274	0.956	0.292	3.271
74°	1.292	0.961	0.276	3.487
75°	1.309	0.966	0.259	3.732
76°	1.326	0.970	0.242	4.011
77°	1.344	0.974	0.225	4.331
78°	1.361	0.978	0.208	4.705
79°	1.379	0.982	0.191	5.145
80°	1.396	0.985	0.174	5.671
81°	1.414	0.988	0.156	6.314
82°	1.431	0.990	0.139	7.115
83°	1.449	0.993	0.122	8.144
84°	1.466	0.995	0.105	9.514
85°	1.484	0.996	0.087	11.43
86°	1.501	0.998	0.070	14.30
87°	1.518	0.999	0.052	19.08
88°	1.536	0.999	0.035	28.64
89°	1.553	1.000	0.018	57.29
90°	1.571	1.000	0.000	∞

Appendix 4

Common Logarithms[a]

N	0	1	2	3	4	5	6	7	8	9
0	....	0000	3010	4771	6021	6990	7782	8451	9031	9542
1	0000	0414	0792	1139	1461	1761	2041	2304	2553	2788
2	3010	3222	3424	3617	3802	3979	4150	4314	4472	4624
3	4771	4914	5051	5185	5315	5441	5563	5682	5798	5911
4	6021	6128	6232	6335	6435	6532	6628	6721	6812	6902
5	6990	7076	7160	7243	7324	7404	7482	7559	7634	7709
6	7782	7853	7924	7993	8062	8129	8195	8261	8325	8388
7	8451	8513	8573	8633	8692	8751	8808	8865	8921	8976
8	9031	9085	9138	9191	9243	9294	9345	9395	9445	9494
9	9542	9590	9638	9685	9731	9777	9823	9868	9912	9956
10	0000	0043	0086	0128	0170	0212	0253	0294	0334	0374
11	0414	0453	0492	0531	0569	0607	0645	0682	0719	0755
12	0792	0828	0864	0899	0934	0969	1004	1038	1072	1106
13	1139	1173	1206	1239	1271	1303	1335	1367	1399	1430
14	1461	1492	1523	1553	1584	1614	1644	1673	1703	1732
15	1761	1790	1818	1847	1875	1903	1931	1959	1987	2014
16	2041	2068	2095	2122	2148	2175	2201	2227	2253	2279
17	2304	2330	2355	2380	2405	2430	2455	2480	2504	2529
18	2553	2577	2601	2625	2648	2672	2695	2718	2742	2765
19	2788	2810	2833	2856	2878	2900	2923	2945	2967	2989
20	3010	3032	3054	3075	3096	3118	3139	3160	3181	3201
21	3222	3243	3263	3284	3304	3324	3345	3365	3385	3404
22	3424	3444	3464	3483	3502	3522	3541	3560	3579	3598
23	3617	3636	3655	3674	3692	3711	3729	3747	3766	3784
24	3802	3820	3838	3856	3874	3892	3909	3927	3945	3962
25	3979	3997	4014	4031	4048	4065	4082	4099	4116	4133
26	4150	4166	4183	4200	4216	4232	4249	4265	4281	4298
27	4314	4330	4346	4362	4378	4393	4409	4425	4440	4456
28	4472	4487	4502	4518	4533	4548	4564	4579	4594	4609
29	4624	4639	4654	4669	4683	4698	4713	4728	4742	4757
30	4771	4786	4800	4814	4829	4843	4857	4871	4886	4900

[a]To obtain Naperian logarithm of a number multiply these logarithms by 2.3026.

Common Logarithms (contd.)

N	0	1	2	3	4	5	6	7	8	9
31	4914	4928	4942	4955	4969	4983	4997	5011	5024	5038
32	5051	5065	5079	5092	5105	5119	5132	5145	5159	5172
33	5185	5198	5211	5224	5237	5250	5263	5276	5289	5302
34	5315	5328	5340	5353	5366	5378	5391	5403	5416	5428
35	5441	5453	5465	5478	5490	5502	5514	5527	5539	5551
36	5563	5575	5587	5599	5611	5623	5635	5647	5658	5670
37	5682	5694	5705	5717	5729	5740	5752	5763	5775	5786
38	5798	5809	5821	5832	5843	5855	5866	5877	5888	5899
39	5911	5922	5933	5944	5955	5966	5977	5988	5999	6010
40	6021	6031	6042	6053	6064	6075	6085	6096	6107	6117
41	6128	6138	6149	6160	6170	6180	6191	6201	6212	6222
42	6232	6243	6253	6263	6274	6284	6294	6304	6314	6325
43	6335	6345	6355	6365	6375	6385	6395	6405	6415	6425
44	6435	6444	6454	6464	6474	6484	6493	6503	6513	6522
45	6532	6542	6551	6561	6571	6580	6590	6599	6609	6618
46	6628	6637	6646	6656	6665	6675	6684	6693	6702	6712
47	6721	6730	6739	6749	6758	6767	6776	6785	6794	6803
48	6812	6821	6830	6839	6848	6857	6866	6875	6884	6893
49	6902	6911	6920	6928	6937	6946	6955	6964	6972	6981
50	6990	6998	7007	7016	7024	7033	7042	7050	7059	7067
51	7076	7084	7093	7101	7110	7118	7126	7135	7143	7152
52	7160	7168	7177	7185	7193	7202	7210	7218	7226	7235
53	7243	7251	7259	7267	7275	7284	7292	7300	7308	7316
54	7324	7332	7340	7348	7356	7364	7372	7380	7388	7396
55	7404	7412	7419	7427	7435	7443	7451	7459	7466	7474
56	7482	7490	7497	7505	7513	7520	7528	7536	7543	7551
57	7559	7566	7574	7582	7589	7597	7604	7612	7619	7627
58	7634	7642	7649	7657	7664	7672	7679	7686	7694	7701
59	7709	7716	7723	7731	7738	7745	7752	7760	7767	7774
60	7782	7789	7796	7803	7810	7818	7825	7832	7839	7846
61	7853	7860	7868	7875	7882	7889	7896	7903	7910	7917
62	7924	7931	7938	7945	7952	7959	7966	7973	7980	7987
63	7993	8000	8007	8014	8021	8028	8035	8041	8048	8055
64	8062	8069	8075	8082	8089	8096	8102	8109	8116	8122
65	8129	8136	8142	8149	8156	8162	8169	8176	8182	8189
66	8195	8202	8209	8215	8222	8228	8235	8241	8248	8254
67	8261	8267	8274	8280	8287	8293	8299	8306	8312	8319
68	8325	8331	8338	8344	8351	8357	8363	8370	8376	8382
69	8388	8395	8401	8407	8414	8420	8426	8432	8439	8445

Common Logarithms (contd.)

N	0	1	2	3	4	5	6	7	8	9
70	8451	8457	8463	8470	8476	8482	8488	8494	8500	8506
71	8513	8519	8525	8531	8537	8543	8549	8555	8561	8567
73	8573	8579	8585	8591	8597	8603	8609	8615	8621	8627
73	8633	8639	8645	8651	8657	8663	8669	8675	8681	8686
74	8692	8698	8704	8710	8716	8722	8727	8733	8739	8745
75	8751	8756	8762	8768	8774	8779	8785	8791	8797	8802
76	8808	8814	8820	8825	8831	8837	8842	8848	8854	8859
77	8865	8871	8876	8882	8887	8893	8899	8904	8910	8915
78	8921	8927	8932	8938	8943	8949	8954	8960	8965	8971
79	8976	8982	8987	8993	8998	9004	9009	9015	9020	9025
80	9031	9036	9042	9047	9053	9058	9063	9069	9074	9079
81	9085	9090	9096	9101	9106	9112	9117	9122	9128	9133
82	9138	9143	9149	9154	9159	9165	9170	9175	9180	9186
83	9191	9196	9201	9206	9212	9217	9222	9227	9232	9238
84	9243	9248	9253	9258	9263	9269	9274	9279	9284	9289
85	9294	9299	9304	9309	9315	9320	9325	9330	9335	9340
86	9345	9350	9355	9360	9365	9370	9375	9380	9385	9390
87	9395	9400	9405	9410	9415	9420	9425	9430	9435	9440
88	9445	9450	9455	9460	9465	9469	9474	9479	9484	9489
89	9494	9499	9504	9509	9513	9518	9523	9528	9533	9538
90	9542	9547	9552	9557	9562	9566	9571	9576	9581	9586
91	9590	9595	9600	9605	9609	9614	9619	9624	9628	9633
92	9638	9643	9647	9652	9657	9661	9666	9671	9675	9680
93	9685	8689	9694	9699	9703	9708	9713	9717	9722	9727
94	9731	9736	9741	9745	9750	9754	9759	9763	9768	9773
95	9777	9782	9786	9791	9795	9800	9805	9809	9814	9818
96	9823	9827	9832	9836	9841	9845	9850	9854	9859	9863
97	9868	9872	9877	9881	9886	9890	9894	9899	9903	9908
98	9912	9917	9921	9926	9930	9934	9939	9943	9948	9952
99	9956	9961	9965	9969	9974	9978	9983	9987	9991	9996
100	0000	0004	0009	0013	0017	0022	0026	0030	0035	0039
N	0	1	2	3	4	5	6	7	8	9

Appendix 5

Masses in Atomic Mass Units for Neutral Atoms of Stable Nuclides and Unstable Nuclides (Designated by *)[a]

Element	Mass Number	Atomic Mass, amu	Element	Mass Number	Atomic Mass, amu
$_0n$	1*	1.008 665		19*	19.001 881
$_1H$	1	1.007 825		20	19.992 440
	2	2.014 102		21	20.993 849
	3*	3.016 050		22	21.991 385
$_2He$	3	3.016 030		23*	22.994 473
	4	4.002 603	$_{11}Na$	22*	21.994 437
	6*	6.018 893		23	22.989 771
$_3Li$	6	6.015 125	$_{12}Mg$	23*	22.994 125
	7	7.016 004		24	23.990 962
	8*	8.022 487		25	24.989 955
$_4Be$	7*	7.016 929		26	25.991 740
	9	9.012 186	$_{13}Al$	27	26.981 539
	10	10.013 534	$_{14}Si$	28	27.976 930
$_5B$	8*	8.024 609		29	28.976 496
	10*	10.012 939		30	29.973 763
	11	11.009 305	$_{15}P$	31	30.973 765
	12*	12.014 354	$_{16}S$	32	31.972 074
$_6C$	10*	10.016 810		33	32.971 462
	11*	11.011 432		34	33.967 865
	12	12.000 000		36	35.967 090
	13	13.003 354	$_{17}Cl$	35	34.968 851
	14*	14.003 242		36*	35.968 309
	15*	15.010 599		37	36.965 898
$_7N$	12*	12.018 641	$_{18}Ar$	36	35.967 544
	13*	13.005 738		38	37.962 728
	14	14.003 074		40	39.962 384
	15	15.000 108	$_{19}K$	39	38.963 710
	16*	16.006 103		40*	39.964 000
	17*	17.008 450		41	40.961 832
$_8O$	14*	14.008 597	$_{20}Ca$	40	39.962 589
	15*	15.003 070		41*	40.962 275
	16	15.994 915		42	41.958 625
	17	16.999 133		43	42.958 780
	18	17.999 160		44	43.955 49
	19*	19.003 578		46	45.953 689
$_9F$	17*	17.002 095		48	47.952 531
	18*	18.000 937	$_{21}Sc$	41*	40.969 247
	19	18.998 405		45	44.955 919
	20*	19.999 987	$_{22}Ti$	46	45.952 632
	21*	20.999 951		47	46.951 769
$_{10}Ne$	18*	18.005 711		48	47.947 951

*Masses in Atomic Mass Units for Neutral Atoms of Stable Nuclides and Unstable Nuclides (Designated by *) Contd.[a]*

Element	Mass Number	Atomic Mass, amu	Element	Mass Number	Atomic Mass, amu
	49	48.947 871	$_{35}$Br	79	78.918 330
	50	49.944 786		81	80.916 292
$_{23}$V	48*	47.952 259	$_{36}$Kr	78	77.920 403
	50*	49.947 164		80	79.916 380
	51	50.943 962		82	81.913 482
$_{24}$Cr	48*	47.953 760		83	82.914 132
	50	49.946 055		84	83.911 504
	52	51.940 514		86	85.910 616
	53	52.940 653	$_{37}$Rb	85	84.911 800
	54	53.938 882		87*	86.909 187
$_{25}$Mn	54*	53.940 362	$_{38}$Sr	84	83.913 431
	55	54.938 051		86	85.909 285
$_{26}$Fe	54	53.939 617		87	86.908 893
	56	55.934 937		88	87.905 641
	57	56.935 398	$_{39}$Y	89	88.905 872
	58	57.933 282	$_{40}$Zr	90	89.904 700
$_{27}$Co	59	58.933 190		91	90.905 642
	60*	59.933 814		92	91.905 031
$_{28}$Ni	58	57.935 342		94	93.906 314
	60	59.930 787		96	95.908 286
	61	60.931 056	$_{41}$Nb	93	92.906 382
	62	61.928 342	$_{42}$Mo	92	91.906 811
	64	63.927 958		94	93.905 091
$_{29}$Cu	63	62.929 592		95	94.905 839
	65	64.927 786		96	95.904 674
$_{30}$Zn	64	63.929 145		97	96.906 022
	66	65.926 052		98	97.905 409
	67	66.927 145		100	99.907 475
	68	67.924 857	$_{44}$Ru	96	95.907 598
	70	69.925 334		98	97.905 289
$_{31}$Ga	69	68.925 574		99	98.905 936
	71	70.924 706		100	99.904 218
$_{32}$Ge	70	69.924 252		101	100.905 577
	72	71.922 082		102	101.904 348
	73	72.923 463		104	103.905 430
	74	73.921 181	$_{45}$Rh	103	102.905 511
	76	75.921 406	$_{46}$Pd	102	101.905 609
$_{33}$As	75	74.921 597		104	103.904 011
$_{34}$Se	74	73.922 476		105	104.905 064
	76	75.919 207		106	105.903 479
	77	76.919 911		108	107.903 891
	78	77.917 314		110	109.905 164
	80	79.916 528	$_{47}$Ag	107	106.905 094
	82	81.916 707		109	108.904 756

*Masses in Atomic Mass Units for Neutral Atoms of Stable Nuclides and Unstable Nuclides (Designated by *) Contd.*[a]

Element	Mass Number	Atomic Mass, amu	Element	Mass Number	Atomic Mass, amu
$_{48}$Cd	106	105.906 463		135	134.905 550
	108	107.904 187		136	135.904 300
	110	109.903 012		137	136.905 500
	111	110.904 189		138	137.905 000
	112	111.902 763	$_{57}$La	138*	137.906 910
	113	112.904 409		139	138.906 140
	114	113.903 361	$_{58}$Ce	126	135.907 100
	116	115.904 762		138	137.905 830
$_{49}$In	113	112.904 089		140	139.905 392
	115*	114.903 871		142	141.909 140
$_{50}$Sn	112	111.904 835	$_{59}$Pr	141	140.907 596
	114	113.902 773	$_{60}$Nd	142	141.907 663
	115	114.903 346		143	142.909 779
	116	115.901 745		144*	143.910 039
	117	116.902 959		145	144.912 538
	118	117.901 606		146	145.913 086
	119	118.903 314		148	147.916 869
	120	119.902 199		150	149.920 915
	122	121.903 442	$_{62}$Sm	144	143.911 989
	124	123.905 272		147*	146.914 867
$_{51}$Sb	121	120.903 817		148	147.914 791
	123	122.904 213		149	148.917 180
$_{52}$Te	120	119.904 023		150	149.917 276
	122	121.903 066		152	151.919 756
	123	122.904 277		154	153.922 282
	124	123.902 842	$_{63}$Eu	151	150.919 838
	125	124.904 418		153	152.921 242
	126	125.903 322	$_{64}$Gd	152	151.919 794
	128	127.904 476		154	153.920 929
	130	129.906 238		155	154.922 664
$_{53}$I	127	126.904 470		156	155.922 175
$_{54}$Xe	124	123.906 120		157	156.924 025
	126	125.904 288		158	157.924 178
	128	127.903 540		160	159.927 115
	129	128.904 784	$_{65}$Tb	159	158.925 351
	130	129.903 509	$_{66}$Dy	156	155.923 930
	131	130.905 086		158	157.924 449
	132	131.904 161		160	159.925 202
	134	133.905 398		161	160.926 945
	136	135.907 221		162	161.926 803
$_{55}$Cs	133	132.905 355		163	162.928 755
$_{56}$Ba	130	129.906 245		164	163.929 200
	132	131.905 120	$_{67}$Ho	165	164.930 421
	134	133.904 612	$_{68}$Er	162	161.928 740

Masses in Atomic Mass Units for Neutral Atoms of Stable Nuclides and Unstable Nuclides (Designated by *) (Contd.)[a]

Element	Mass Number	Atomic Mass, amu	Element	Mass Number	Atomic Mass, amu
	164	163.929 287		187	186.955 832
	166	165.930 307		188	187.956 081
	167	166.932 060		189	188.958 300
	168	167.932 383		190	189.958 630
	170	169.935 560		192	191.961 450
$_{69}$Tm	169	168.934 245	$_{77}$Ir	191	190.960 640
$_{70}$Yb	168	167.934 160		193	192.963 012
	170	169.935 020	$_{78}$Pt	190*	189.959 950
	171	170.936 430		192	191.961 150
	172	171.936 360		194	193.962 725
	173	172.938 060		195	194.964 813
	174	173.938 740		196	195.964 967
	176	175.942 680		198	197.967 895
$_{71}$Lu	175	174.940 640	$_{79}$Au	197	196.966 541
	176*	175.942 660	$_{80}$Hg	196	195.965 820
$_{72}$Hf	174	173.940 360		198	197.966 756
	176	175.941 570		199	198.968 279
	177	176.943 400		200	199.968 327
	178	177.943 880		201	200.970 308
	179	178.946 030		202	201.970 642
	180	179.946 820		204	203.973 495
$_{73}$Ta	181	180.948 007	$_{81}$Ti	203	202.972 353
$_{74}$W	180	179.947 000		205	204.974 442
	182	181.948 301	$_{82}$Pb	204	203.973 044
	183	182.950 324		206	205.974 468
	184	183.951 025		207	206.975 903
	186	185.954 440		208	207.976 650
$_{75}$Re	185	184.953 059	$_{83}$Bi	209	208.981 082
	187*	186.955 833	$_{90}$Th	232*	232.038 124
$_{76}$Os	184	183.952 750	$_{92}$U	234*	234.040 904
	186	185.953 870		235*	235.043 915
	188	187.956 081		238*	238.050 770

[a]From R. Weidner and R. Sells. 1968. Elementary modern physics, 2nd Ed. Allyn and Bacon, Boston.

Answers to Problems

Chapter 1

1. 34,750 eV
   $-4,140$ eV
2. Isotopes: $^{44}_{22}$Ti, $^{45}_{22}$Ti, $^{46}_{22}$Ti, $^{47}_{22}$Ti; $^{45}_{23}$V, $^{46}_{23}$V
   Isotones: $^{45}_{21}$Sc, $^{46}_{22}$Ti; $^{44}_{22}$Ti, $^{45}_{23}$V, $^{46}_{24}$Cr
   Isobars: $^{45}_{20}$Ca, $^{45}_{21}$Sc, $^{45}_{22}$Ti, $^{45}_{23}$V; $^{46}_{22}$Ti, $^{46}_{23}$V, $^{46}_{24}$Cr
3. 0.1369 amu
   7.97 MeV/nucleon
   $Z = 8; A = 16$
4. 200 MeV
5. 17.6 MeV
6. 24×10^{23}
   0.85 g

Chapter 2

1. 12.0 hr
   18.0 hr
   134 hr
2. 31 g
   10.4×10^{17} atoms
3. $^{210}_{83}$Bi $\rightarrow$ $^{206}_{81}$Tl $+ ^4_2$He
   $^{210}_{83}$Bi $\rightarrow$ $^{210}_{84}$Po $+ ^{\ 0}_{-1}\beta + \nu$
4. 6×10^{19}/sec
   0.05 Å
5. $^{108}_{47}$Ag $\rightarrow$ $^{108}_{48}$Cd $+ ^{\ 0}_{-1}\beta + \nu$
   $^{108}_{47}$Ag $\rightarrow$ $^{108}_{46}$Pd $+^{\ 0}_{+1}\beta + \nu$
   $^{108}_{47}$Ag $+ ^{\ 0}_{-1}$e $\rightarrow$ $^{108}_{46}$Pd $+ \nu$
6. (a) unstable
   (b) positron
   (c) $^{12}_7$N
   (d) $^{12}_5$B
   (e) 19.4 sec
   (f) 98.89%
   (g) 13.003354
   (h) 5.3 MeV
   (i) 1.87 MeV
7. 6.2×10^{14}
   0.093 μg
8. ^{18}F: ec and $^{\ 0}_{+1}\beta$
   ^{20}F: $^{\ 0}_{-1}\beta$
9. 75
10. 95 mCi

Chapter 3

1. 4.4 keV/cm
2. 9.3×10^{-4}
3. $135 \ gm/cm^2$
   $6.45 \ mg/cm^2$
4. 0.65 MeV
5. $200 \ mg/cm^2$

Chapter 4

1. 32.9 keV; 24.6 keV
2. 537 keV; 197 keV; 125 keV; 465 keV
3. 0.87 MeV
4. 0.57 cm; 1.22/cm
5. 0.35
6. 1/1024
7. 0; 0.80 MeV

Chapter 5

1. 67 mCi
2. (a) $^{72}_{34}Se$
   (b) $^{103}_{44}Ru$
   (c) n
   (d) α
   (e) α
   (f) $^{33}_{15}P$
   (g) p
   (h) $^{14}_{7}N$
   (i) n
   (j) d
   (k) 2n; $^{237}_{93}Np$
   (l) p
3. $^{32}_{15}P$; $^{29}_{14}Si$; $^{31}_{15}P$; $^{33}_{16}S$
4. 3.4×10^{-10}
5. 121 barns

Chapter 6

1. 0.04%
2. 26%

Chapter 8

1. 0.34 mV
2. 2×10^9
3. 68 ml
4. (a) no
   (b) yes; yes
   (c) yes
5. 0.55(counts/neutron-cm²)

Chapter 9

1. 6.6×10^6

Chapter 11

1. 17 sec

Chapter 12

1. 692 V
4. 3.13 MeV = photopeak
   2.62 MeV = single escape peak
   2.11 MeV = double escape peak
   0.51 MeV = annihilation peak
   0.20 MeV = backscatter peak
5. Sum peak
6. X-Ray escape peak

Chapter 16

1. 0.051 μCi
2. 0.00093 μCi

Chapter 18

1. (a) 5.7/min; 1.8%
   (b) 3.6/min; 3.5%
   (c) 6.7/min; 3.0%
2. $p = 0.124$; no
3. 20%; 6%
4. yes; 175

Chapter 19

1. 2.4×10^{10}
2. 8270/min; 2400/min
3. 64%; 1.64
5. 74/min/mg
7. (a) 108 (mCi/ml);
   (b) 0.044

Chapter 20

2. 2720 ml
2. 27%
4. 55%; abnormally high

Chapter 22

1. 1.05×10^{11}
2. 11.5 R
3. 37.1 rem
4. 360 rad
5. 180 mR/hr
6. 0.175 (R-m²/hr-Ci)

Chapter 23

1. 0.12 μrad/sec; 6.0 mrad; 8.8 mrad
2. 0.24 μrad/sec; 12.6 mrad; 18.5 mrad
3. 0.36 μrad/sec; 18.6 mrad; 27.3 mrad
4. 22.2 days; 29.8 days

Index